SCHAUM'S OUTLINE OF

THEORY AND PROBLEMS

of

EARTH
SCIENCES

•

by

ARTHUR BEISER, Ph.D.

SCHAUM'S OUTLINE SERIES
McGRAW-HILL BOOK COMPANY

New York   St. Louis   San Francisco   Auckland   Düsseldorf   Johannesburg
Kuala Lumpur   London   Mexico   Montreal   New Delhi   Panama
Paris   São Paulo   Singapore   Sydney   Tokyo   Toronto

07-004375-2

1 2 3 4 5 6 7 8 9 10 11 12 13 14 15 16 17 18 19 20   SH SH   7 9 8 7 6 5

Library of Congress Cataloging in Publication Data

Beiser, Arthur.
   Schaum's outline of theory and problems of earth sciences.
   (Schaum's outline series)
   1. Earth sciences—Outlines, syllabi, etc.
I. Title.   II. Title: Earth sciences.
[QE41.B385]   550   75-34139
ISBN 0-07-004375-2

Preface

This book is intended to provide students of earth sciences with help in mastering a wide spectrum of topics, including geology, astronomy, mineralogy, weather, matter and energy. Both SI (metric) and British units are used.

Each chapter begins with an outline of its subject. The solved problems that follow are of two kinds: those that show how numerical answers are obtained to typical questions in physics and chemistry, and those that review important facts and ideas in all the physical sciences. The supplementary problems give the reader both a chance for practice and a means to gauge his progress.

ARTHUR BEISER

CONTENTS

CONTENTS

CONTENTS

Chapter 1

Matter and Energy

ENERGY

Work is a measure of the amount of change (in a general sense) a force produces when it acts upon a body. The change may be in the velocity of the body, in its position, in its size or shape, and so forth. The work done by a force acting on a body is equal to the product of the force and the distance through which the force acts, provided the force and the displacement are in the same direction. In the British system of units, the unit of work is the *foot-pound,* which is the work done by a force of one pound that acts through a distance of one foot. In the International System of Units (SI), the unit of work is the *joule* (J), where 1 J = 0.738 ft-lb.

Energy is that property something has which enables it to do work. The more energy something has, the more work it can perform. Every kind of energy falls into one of three general categories: kinetic energy, potential energy, and rest energy. The units of energy are the same as those of work, namely the foot-pound and the joule.

CONSERVATION OF ENERGY

The energy a body has by virtue of its motion is called *kinetic energy*. The greater the mass of the body and the faster it is moving, the greater its kinetic energy is.

The energy a body has by virtue of its position is called *potential energy*. A book held above the floor has gravitational potential energy because the book can do work on something else as it falls; the higher the book, the more potential energy it has. A nail held near a magnet has magnetic potential energy because the nail can do work as it moves toward the magnet, and the wound spring in a watch has elastic potential energy because the spring can do work as it unwinds.

Matter can be converted into energy, and energy into matter. The *rest energy* of a body is the energy it has by virtue of its mass alone. Thus mass can be regarded as a form of energy. The rest energy of a body is in addition to any kinetic or potential energy it might also have.

According to the law of *conservation of energy*, energy cannot be created or destroyed, although it can be transformed from one kind into another. The total amount of energy in the universe is constant. A falling stone provides a simple example: more and more of its initial potential energy turns into kinetic energy as its velocity increases, until finally all the potential energy has become kinetic energy when it strikes the ground. The kinetic energy of the stone is then transferred to the ground by the impact.

ATOMS AND MOLECULES

Elements are the fundamental substances of which all matter in bulk is composed. There are 105 known elements, of which a number are not found in nature but have been prepared in the laboratory. Elements cannot be transformed into one another by ordinary chemical or physical means, but two or more elements can combine to form a *compound*, which is a substance whose properties are different from those of its constituent elements.

The ultimate particles of an element are called *atoms,* and those of a compound which exists in the gaseous state are called *molecules*. The molecules of a compound consist of the atoms of the elements that compose it joined together in a specific arrangement; each molecule of water, for instance, contains two hydrogen atoms and one oxygen atom, as its symbol H_2O indicates. Many compounds in the solid and liquid states do not consist of individual molecules, as discussed below and in Chapter 5.   Elemental gases may consist of atoms (helium, He; argon, Ar) or of molecules (hydrogen, H_2; oxygen, O).

ATOMIC STRUCTURE

An atom of any element consists of a small *nucleus* with a number of *electrons* some distance away.   The nucleus is composed of *protons*, which have positive electric charges, and *neutrons*, which are uncharged.   Electrons are negatively charged and are much lighter than protons and neutrons, so nearly all the mass of an atom resides in its nucleus.   The number of protons in the nucleus of an atom is normally equal to the number of electrons around it, so that the atom as a whole is electrically neutral.   The forces between atoms that hold them together as molecules, solids, and liquids are electrical in origin.

IONS

Under certain circumstances an atom may lose one or more electrons and become a *positive ion,* or it may gain one or more electrons and become a *negative ion*.   Many solids consist of positive and negative ions rather than of atoms or molecules.   An example is ordinary table salt, which is made up of positive sodium ions (Na^+) and negative chlorine ions (Cl^-).   Solutions of such solids in water also contain ions.   Sparks, flames, and X-rays are among the influences that can ionize gases.   Ions of opposite sign in a gas come together soon after being formed and the excess electrons on the negative ions pass to the positive ones to form neutral molecules.   A gas can be maintained in an ionized state by passing an electric current through it (as in a neon sign) or by bombarding it with X-rays or ultraviolet light (as in the upper atmosphere of the earth, where the radiation comes from the sun).

INTERNAL ENERGY

Every body of matter, whether solid, liquid, or gas, consists of atoms or molecules which are in rapid motion.   The kinetic energies of these particles constitute the *internal energy* of the body of matter.   The *temperature* of the body is a measure of the average kinetic energy of its particles. *Heat* may be thought of as internal energy in transit.   When heat is added to a body, its internal energy increases and its temperature rises; when heat is removed from a body, its internal energy decreases and its temperature falls.

Temperature is familiar as the property of a body of matter responsible for sensations of hot or cold when it is touched.   Temperature provides an indicator of the direction of internal flow: when two objects are in contact, internal energy goes from the one at the higher temperature to the one at the lower temperature, regardless of the total amounts of internal energy in each one.   Thus if hot coffee is poured into a cold cup, the coffee becomes cooler and the cup becomes warmer.

Because heat is a form of energy, the proper SI unit of heat is the joule.   However, the *kilocalorie* is still widely used with SI units: 1 kilocalorie (kcal) is the amount of heat needed to raise the temperature of 1 kg of water by 1 °C.   The *calorie* itself is the amount of heat needed to raise the temperature of 1 g of water by 1 °C; hence 1 kcal = 1000 calories.   (The calorie used by dieticians to measure the energy content of foods is the same as the kilocalorie.)   For conversion of units: 1 kcal = 4184 J.

CHANGE OF STATE

When heat is continuously added to a solid, it grows hotter and hotter and finally begins to melt.   While it is melting, the material remains at the same temperature and the absorbed heat goes into changing its state from solid to liquid.   After all the solid is melted, the temperature of the resulting liquid then increases as more heat is supplied until it begins to boil.   Now the material again stays at a constant temperature until all of it has become a gas, after which the gas temperature rises.

The amount of heat that must be added to 1 kg of a substance at its melting point to change it from a solid to a liquid is called its *heat of fusion*. The same amount of heat must be removed from the substance when it is a liquid at its melting point to change it to a solid.

The amount of heat that must be added to 1 kg of a substance at its boiling point to change it from a liquid to a gas is called its *heat of vaporization*. The same amount of heat must be removed from the substance when it is a gas at its boiling point to change it to a liquid.

HEAT TRANSFER

The three mechanisms by which heat can be transferred from one place to another are conduction, convection, and radiation.

In *conduction*, heat is carried by means of collisions between rapidly moving molecules at the hot end of a body of matter and the slower molecules at the cold end. Some of the kinetic energy of the fast molecules passes to the slow molecules, and the result of successive collisions is a flow of heat through the body of matter. Solids, liquids, and gases all conduct heat. Conduction is poorest in gases because their molecules are relatively far apart and so interact less frequently than in the case of solids and liquids. Metals are the best conductors of heat because some of their electrons are able to move about relatively freely and can travel past many atoms between collisions.

In *convection*, a volume of hot fluid (gas or liquid) moves from one region to another carrying internal energy with it. When a pan of water is heated on a stove, for instance, the hot water at the bottom expands slightly so that its density decreases, and the buoyancy of this water causes it to rise to the surface while colder, denser water descends to take its place at the bottom.

In *radiation*, energy is carried by the *electromagnetic waves* emitted by every object. Electromagnetic waves, of which light, radio waves, and X-rays are examples, travel at the velocity of light (3×10^8 m/s = 186,000 mi/s) and require no material medium for their passage. The higher the temperature of an object, the greater the rate at which it radiates energy.

POWERS OF TEN

Very small and very large numbers are common in physical science and are best expressed with the help of powers of 10. Any number in decimal form can be written as a number between 1 and 10 multiplied by a power of 10:

$$834 = 8.34 \times 10^2 \qquad 0.00072 = 7.2 \times 10^{-4}$$

The powers of 10 from 10^{-6} to 10^6 are as follows:

10^0	= 1	= 1 with decimal point moved 0 places
10^{-1}	= 0.1	= 1 with decimal point moved 1 place to the left
10^{-2}	= 0.01	= 1 with decimal point moved 2 places to the left
10^{-3}	= 0.001	= 1 with decimal point moved 3 places to the left
10^{-4}	= 0.0001	= 1 with decimal point moved 4 places to the left
10^{-5}	= 0.00001	= 1 with decimal point moved 5 places to the left
10^{-6}	= 0.000001	= 1 with decimal point moved 6 places to the left
10^0	= 1	= 1 with decimal point moved 0 places
10^1	= 10	= 1 with decimal point moved 1 place to the right
10^2	= 100	= 1 with decimal point moved 2 places to the right
10^3	= 1000	= 1 with decimal point moved 3 places to the right
10^4	= 10,000	= 1 with decimal point moved 4 places to the right
10^5	= 100,000	= 1 with decimal point moved 5 places to the right
10^6	= 1,000,000	= 1 with decimal point moved 6 places to the right

Solved Problems

1.1.   What is the ultimate source of the energy in food, coal, oil, natural gas, and falling water?

Sunlight transfers energy from the sun to the earth which becomes stored in chemical compounds in plants through the process of photosynthesis. Animals in turn acquire energy by eating plants (or other animals that eat plants). Food consists of plant and animal tissues, and the energy it contains therefore comes from the sun. Coal, oil, and natural gas are formed from the remains of plants and animals, and the energy they contain can thus also be traced back to the sun. Water is evaporated from the earth's surface by energy supplied by sunlight, and some of it falls as rain and snow on high ground. The increased potential energy of the elevated water turns into kinetic energy as the water flows downhill, and this energy can be extracted by waterwheels and turbines.

1.2.   What is the nature of chemical energy?

Chemical energy is rest energy. When a chemical reaction occurs in which energy is given off, for instance, the products of the reaction always have less mass than the original substances did. Because even a small quantity of matter is equivalent to a vast amount of energy ($1 \text{ kg} = 9 \times 10^{16} \text{ J}$), the mass changes in chemical reactions are too small to be detectable by ordinary means.

1.3.   An iron atom has 26 protons in its nucleus. (*a*) How many electrons does this atom contain? (*b*) How many electrons does the Fe^{+++} ion contain?

(*a*)   Since a normal atom is electrically neutral, the number of negatively charged electrons it contains equals the number of positively charged protons in its nucleus. The iron atom therefore contains 26 electrons.

(*b*)   The symbol Fe^{+++} represents an iron atom that has a net charge of $+3$, which means that it has lost three of its usual complement of electrons. The Fe^{+++} ion therefore contains 23 electrons.

1.4.   What are the four *fundamental forces* that are responsible for all the physical processes in the universe?

The four fundamental forces are gravitational, electromagnetic, strong nuclear, and weak nuclear. The two types of nuclear force have very short ranges and act within atomic nuclei; it is possible that the weak force is related to the electromagnetic force. Electromagnetic forces, which are unlimited in range, act between electrically charged particles and determine the structures of atoms, molecules, solids, and liquids; when one object is in contact with another object, electromagnetic forces are ultimately responsible for the forces they exert on each other. Gravitational forces, also unlimited in range, act between all masses and are important in determining the structures of planets, stars, and galaxies of stars.

1.5.   A glass of water is stirred and then allowed to stand until the water stops moving. What has happened to the kinetic energy of the moving water?

The kinetic energy of the moving water becomes dissipated into internal energy, and the water therefore has a higher temperature than before it was stirred.

1.6.   The *Celsius* (or *centigrade*) temperature scale assigns 0° to the freezing point of water and 100° to its boiling point. On the *Fahrenheit* scale these points are respectively 32° and 212°. A Fahrenheit degree is therefore 5/9 as large as a Celsius degree. The following formulas give the procedure for converting a temperature expressed in one scale to the corresponding value of the other:

$$T_F = \frac{9}{5} T_C + 32°$$

$$T_C = \frac{5}{9} (T_F - 32°)$$

What is the Celsius equivalent of 80 °F?

$$T_C = \frac{5}{9} (T_F - 32°) = \frac{5}{9} (80° - 32°) = 26.7 \text{ °C}$$

1.7.   What is the Fahrenheit equivalent of 80 °C?

$$T_F = \frac{9}{5}T_C + 32° = \frac{9}{5} \times 80° + 32° = 176 \text{ °F}$$

1.8.   Oxygen freezes at -362 °F. What is the Celsius equivalent of this temperature?

$$T_C = \frac{5}{9}(T_F - 32°) = \frac{5}{9}(-362° - 32°) = -219 \text{ °C}$$

1.9.   Nitrogen freezes at -210 °C. What is the Fahrenheit equivalent of this temperature?

$$T_F = \frac{9}{5}T_C + 32° = \frac{9}{5}(-210°) + 32° = -346 \text{ °F}$$

1.10.   Do other substances respond to the addition or removal of a given amount of heat by the same temperature change as the same mass of water would?

Different substances respond differently to the addition or removal of heat. For instance, 1 kg of water increases in temperature by 1 °C when 1 kcal of heat is added, but 1 kg of aluminum increases in temperature by 4.5 °C when this is done. The *specific heat capacity* of a substance is the amount of heat needed to change the temperature of 1 kg of it by 1 °C. Among common materials, water has the highest specific heat capacity. Ice and steam have specific heat capacities about half that of water. Metals usually have low specific heat capacities; that of lead, for example, is only 3% that of water.

1.11.   A person is dissatisfied with the rate at which eggs cook in a pan of boiling water. Would they cook faster if he turns up the gas flame?

No. The maximum temperature that water can have while in the liquid state is its boiling point. Increasing the rate at which the heat is supplied to a pan of water increases the rate at which steam is produced, but does not raise the temperature of the water beyond 100 °C (212 °F).

1.12.   Heat is added at a constant rate to a certain mass of water that is initially ice at -50 °C. Draw a graph showing how the temperature of the water changes as time goes on; indicate where the ice melts into water and where the water boils into steam.

The graph is shown in Fig. 1-1. At first the temperature of the ice increases as heat is supplied to it. When the temperature of the ice reaches 0 °C it begins to melt, and the temperature remains at 0 °C until all the ice has turned into water. Then the temperature of the water rises to 100 °C, when boiling begins to occur. The rate of rise of the water temperature is slower than in the case of the ice because water requires more heat for a given change in temperature than an equal mass of ice does. The temperature remains at 100 °C until the water has all become steam, whereupon the temperature of the steam begins to increase.

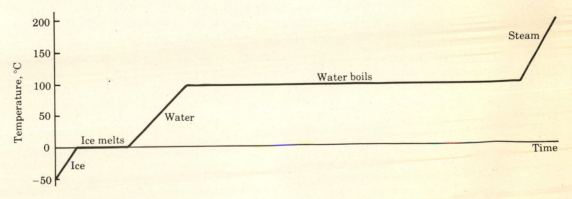

Fig. 1-1

1.13. When a sample of a gas is allowed to expand at a constant temperature, its pressure drops. Why?

A gas is composed of molecules that are in constant random motion. The pressure a gas exerts is due to the impacts of its molecules; there are so many molecules in even a small gas sample that the individual blows appear as a continuous force. Expanding a gas sample means that its molecules must travel farther between successive impacts on the container walls and that the impacts are spread over a larger area (Fig. 1-2). Hence an increase in volume means a decrease in pressure, and vice versa.

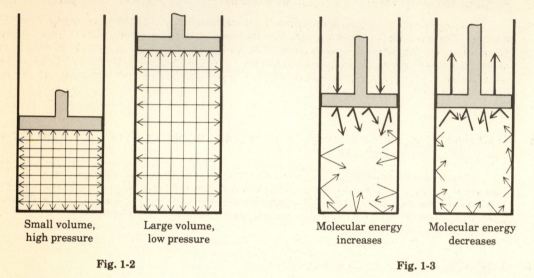

| Small volume, | Large volume, | | Molecular energy | Molecular energy |
| high pressure | low pressure | | increases | decreases |

Fig. 1-2   **Fig. 1-3**

1.14. (*a*) What is the significance of the temperature of a gas in terms of the motions of its molecules? (*b*) When a sample of a gas is compressed, its temperature rises; when the sample is expanded, its temperature falls. Explain these observations.

(*a*) The temperature of a gas is a measure of the average kinetic energy of its molecules. The faster the molecules move, the higher the temperature.

(*b*) Compressing a gas causes its temperature to rise because molecules rebound from the inward-moving walls of the container with increased energy, just as a tennis ball rebounds with greater energy when struck by a racket. Similarly, expanding a gas causes its temperature to fall because molecules rebound from the outward-moving walls with decreased energy (Fig. 1-3). From a macroscopic point of view, we note that work is done on a gas when it is compressed, and the increased energy is manifested in a rise in temperature. An expanding gas does work on its container, and the corresponding loss of energy is manifested in a drop in temperature.

1.15. Gas molecules have velocities comparable with those of rifle bullets, yet we all know that a gas with a strong odor, such as ammonia, takes several seconds to diffuse through a room. Why?

Gas molecules collide frequently with one another, which means that a particular molecule follows a long, very complicated path in going from one place to another.

1.16. What is the origin of the heat of fusion of a solid? Of the heat of vaporization of a liquid?

The molecules of a solid are close enough together to exert forces on one another that hold the entire assembly to a definite size and shape. As in the case of a gas, the molecules are in constant motion, but they vibrate about fixed locations instead of moving randomly. The molecules of a liquid continually move around past one another more or less freely, which enables the liquid to flow, but their spacing does not change and so the volume of a given liquid sample does not vary.

When a solid melts, the original arrangement of its molecules changes to the random arrangement of molecules in a liquid. To accomplish the change, the molecules must be pulled apart against the forces holding them in place, which requires energy. The heat of fusion of a solid represents this energy. When a liquid boils, the heat of vaporization represents the energy needed to pull its molecules entirely free of one another so that a gas is formed.

1.17. Explain the evaporation of a liquid at a temperature below its boiling point.

At any moment in a liquid, some molecules are moving faster and others are moving slower than the average. The fastest ones are able to escape from the liquid surface despite the attractive forces exerted by the other molecules; this constitutes evaporation. The warmer the liquid, the greater the number of very fast molecules, and the more rapidly evaporation takes place. Since the molecules that remain behind are the slower ones, the liquid has a lower temperature than before (unless heat has been added to it from an outside source during the process).

1.18. If all objects radiate electromagnetic energy, why do not the objects around us in everyday life grow colder and colder?

Every object also absorbs electromagnetic energy from its surroundings, and if both object and surroundings are at the same temperature, energy is emitted and absorbed at the same rate. When an object is at a higher temperature than its surroundings and heat is not supplied to it, it radiates more energy than it absorbs and cools down to the temperature of its surroundings.

1.19. Examples of powers-of-10 notation.

$$20 = 2 \times 10 = 2 \times 10^1$$
$$3043 = 3.043 \times 1000 = 3.043 \times 10^3$$
$$8,700,000 = 8.7 \times 1,000,000 = 8.7 \times 10^6$$
$$0.22 = 2.2 \times 0.1 = 2.2 \times 10^{-1}$$
$$0.000035 = 3.5 \times 0.00001 = 3.5 \times 10^{-5}$$

Supplementary Problems

1.20. (a) Do all moving bodies possess kinetic energy? (b) Do all stationary bodies possess potential energy? (c) Is it possible for a body to possess both kinetic and potential energy?

1.21. Can you suggest any sources of energy found on the earth that cannot be traced back to the sun's radiation?

1.22. Why does a nail become hot when it is hammered into a piece of wood?

1.23. Why is an ice cube at 0 °C more effective in cooling a drink than the same mass of water at 0 °C?

1.24. Ethyl alcohol melts at −114 °C and boils at 78 °C. What are the Fahrenheit equivalents of these temperatures?

1.25. Bromine melts at 19 °F and boils at 140 °F. What are the Celsius equivalents of these temperatures?

1.26. The symbol for the normal oxygen atom is O. What is the symbol for an oxygen atom that has acquired two additional electrons? Is this ion positively or negatively charged?

1.27. The temperature of a gas sample is raised. Why does the pressure the gas exerts also increase?

1.28. How does perspiration give the body a means of cooling itself?

1.29. Outdoors in the winter, why does a piece of metal feel colder than a piece of wood?

1.30. Express the following numbers in powers-of-10 notation:

(a) 720   (d) 0.000062   (g) 49,527   (j) 49,000,000,000
(b) 890,000   (e) 3.6   (h) 0.002943   (k) 0.000000011
(c) 0.02   (f) 0.4   (i) 0.0014   (l) 1.4763

1.31. Express the following numbers in decimal notation:

(a) 3×10^{-4} (c) 8.126×10^{-5} (e) 5×10^2 (g) 4.32145×10^3 (i) 5.7×10^0
(b) 7.5×10^3 (d) 1.01×10^8 (f) 3.2×10^{-2} (h) 6×10^6 (j) 6.9×10^{-5}

Answers to Supplementary Problems

1.20. Yes; no; yes.

1.21. Nuclear energy, which can be released in nuclear reactors and weapons; geothermal energy, which is the energy of the earth's heat and largely originates in radioactivity in the interior of the earth.

1.22. Most of the work done in hammering the nail is dissipated as heat owing to friction between the nail and the wood.

1.23. The ice absorbs heat from the drink in order to melt to water at 0 °C.

1.24. −173 °F; 172 °F.

1.25. −7 °C; 60 °C.

1.26. Since electrons are negatively charged, the ion is also negatively charged, and its symbol is O^{--}.

1.27. The pressure increases because the gas molecules move faster than before and therefore both strike the walls more often and produce a greater force with each impact.

1.28. Perspiration provides a liquid to be evaporated from the skin, with the required heat coming from the body.

1.29. Metals are much better conductors of heat than wood and therefore conduct heat away from the hand more rapidly.

1.30.

(a) 7.2×10^2 (g) 4.9527×10^4
(b) 8.9×10^5 (h) 2.943×10^{-3}
(c) 2×10^{-2} (i) 1.4×10^{-3}
(d) 6.2×10^{-5} (j) 4.9×10^{10}
(e) 3.6×10^0 (k) 1.1×10^{-8}
(f) 4×10^{-1} (l) 1.4763×10^0

1.31.

(a) 0.0003   (f) 0.032
(b) 7500   (g) 4321.45
(c) 0.00008126   (h) 6,000,000
(d) 101,000,000   (i) 5.7
(e) 500   (j) 0.000069

Chapter 2

The Atmosphere

COMPOSITION

The earth's atmosphere consists chiefly of nitrogen (78% by volume) and oxygen (21%). The remainder is largely argon (0.9%) and carbon dioxide (0.03%), plus traces of a number of other gases. Water vapor is present as well but to a variable extent, ranging from nearly 0 to 4%. The lower atmosphere also contains a considerable quantity of small, solid particles of different kinds (such as soot, bits of rock and soil, salt grains from the evaporation of sea water, and spores, pollen, and bacteria); these particles provide nuclei for the condensation of atmospheric water vapor to form clouds, fog, rain, and snow.

Nitrogen, oxygen, and carbon dioxide are important biologically. Nitrogen is a key ingredient of the amino acids of which all proteins consist, and certain bacteria are able to convert atmospheric nitrogen into nitrogen compounds which plants can utilize in manufacturing amino acids. Plants also convert water and atmospheric CO_2 into carbohydrates and oxygen in photosynthesis; animals obtain the carbohydrates and amino acids they need by eating plants. Plants and animals both derive energy by using atmospheric oxygen to convert carbon in their foods to CO_2.

High in the atmosphere solar X- and ultraviolet radiation split N_2 and O_2 molecules into atoms and into ions and electrons. One result is the formation of a small amount of *ozone*, O_3, from reactions between O_2 molecules and O atoms. Ozone is a very efficient absorber of solar ultraviolet at wavelengths longer than those absorbed by N_2 and O_2 and so prevents this potentially lethal radiation from reaching the earth's surface. The region of the upper atmosphere that contains ions and electrons is known as the *ionosphere*; long-range radio communication is possible because radio waves are channeled between the earth's surface and the ionosphere by reflection at both, instead of simply escaping into space.

STRUCTURE

The character of the atmosphere changes more or less abruptly at the *tropopause, stratopause,* and *mesopause,* which occur respectively at altitudes that average about 10 km (6 mi), 50 km (31 mi), and 80 km (50 mi). These surfaces divide the atmosphere into four regions, listed in order of increasing altitude (Fig. 2-1):

1. The *troposphere* is the dense lower part of the atmosphere in which meteorological phenomena such as clouds and storms occur. Air temperature in the troposphere decreases with altitude until it reaches about −55 °C at the tropopause.

2. In the *stratosphere* the air is clear and dry. The temperature is constant in the lower part of the stratosphere but then rises because of heating from the absorption of solar energy by the ozone layer. The stratopause is defined by a temperature maximum, which is usually about 10 °C.

3. In the *mesosphere* the temperature falls steadily to about −80 °C at the mesopause. Air pressure at the mesopause is only about 3/100,000 of sea-level pressure.

4. In the *thermosphere* the absorption of solar X- and ultraviolet radiation results in high temperatures (1000 °C or more) and considerable ionization. The various layers of the ionosphere are found here. It must be kept in mind that the high temperatures of the thermosphere represent the average molecular energies there; because the number of

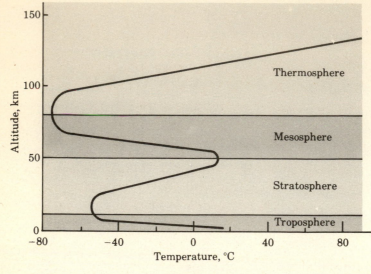

Fig. 2-1

molecules per m³ is so small, the total energy per m³ is also small despite the high temperature.

ENERGY BALANCE

Solar radiation reaches the top of the atmosphere at the rate of 203 kcal/min per m² of area perpendicular to its direction; about one-third of the arriving energy is reflected back into space, largely by clouds. Since the average temperature of the earth's surface does not change when reckoned on a long-term basis, the earth and its atmosphere reradiate away as much energy as they absorb. However, although intake and outgo are in balance for the earth as a whole, tropical regions receive more direct solar energy than they radiate, and polar regions receive less. This is shown in Fig. 2-2; the latitude scale in the graph is spaced so that equal horizontal distances correspond to equal areas of the earth's surface. Large-scale winds in the atmosphere and, to a lesser extent, currents in the oceans shift energy from the tropics to the high latitudes. The atmosphere receives the energy it carries around the globe from long-wavelength infrared radiation given off by the earth, which is absorbed by carbon dioxide and water vapor and communicated to the other atmospheric gases in molecular collisions.

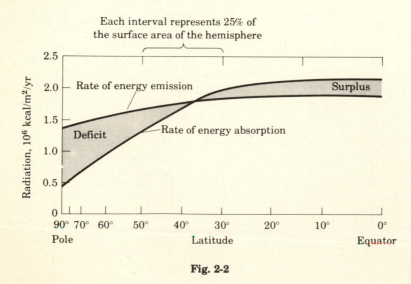

Fig. 2-2

MOISTURE

The atmosphere transports water as well as energy around the world. More water falls as rain and snow on land masses than is lost by them through evaporation, which compensates for the runoff of continental water to the oceans by rivers and streams; in turn the oceans lose more water by evaporation than they gain by precipitation.

Air which contains the maximum amount of water vapor that can evaporate at a given temperature is said to be *saturated*. The higher the temperature, the greater the concentration of water vapor at saturation. The *relative humidity* of a volume of air refers to its degree of saturation: relative humidities of 0, 50%, and 100% mean respectively that there is no moisture present, that the air contains half as much moisture as the maximum possible, and that the air is saturated. If a volume of air is cooled past the temperature at which it is saturated, the excess water vapor will condense on suitable nuclei into droplets of water or, in certain cirumstances, into ice crystals.

Clouds form when rising air is cooled by its expansion. The water vapor usually condenses around salt particles in the air and forms tiny droplets or ice crystals small enough to remain suspended aloft indefinitely. Precipitation occurs when some of the water droplets or ice crystals in a cloud coalesce with others and grow heavy enough to fall.

Solved Problems

2.1. In what two important processes does the carbon dioxide in the atmosphere have essential roles?

    (a) In photosynthesis, plants manufacture carbohydrates from atmospheric CO_2 and water, with oxygen as a by-product.

    (b) By absorbing infrared radiation emitted by the earth, CO_2 is an intermediary in the process by which solar energy is transferred to the lower atmosphere and carried around the earth by winds.

2.2. Illustrate the nitrogen cycle with a diagram.

    See Fig. 2-3.

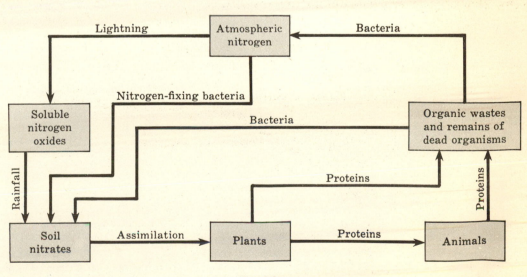

Fig. 2-3

2.3.   Illustrate the oxygen–carbon dioxide cycle with a diagram.

See Fig. 2-4.

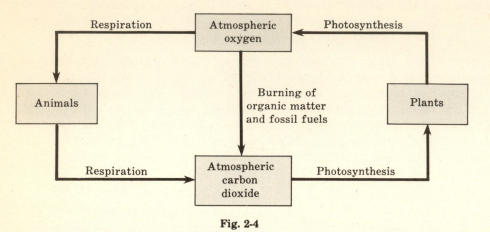

Fig. 2-4

2.4.   What are the five chief classes of atmospheric pollutants?

Suspended particles (smoke, dust); carbon monoxide; unburned hydrocarbons; sulfur oxides (SO_2, SO_3); and nitrogen oxides (NO, NO_2).

2.5.   (*a*) What is ozone?   (*b*) Why is the ozone content of the atmosphere concentrated in a layer rather than being distributed uniformly at all levels?   (*c*) Why is the ozone layer so important for terrestrial life?

(*a*)   Ozone is a form of oxygen whose molecules consist of three O atoms, so that its formula is O_3.   Ozone is less stable than O_2 and tends to break up into O_2 + O.

(*b*)   The first step in the formation of an ozone molecule is the decomposition of an O_2 molecule into two O atoms. The energy required for this process is provided by photons of ultraviolet light from the sun.   The second step is the attachment of an O atom to an O_2 molecule to form O_3.   The rate of ozone production thus depends upon both the O_2 concentration and the intensity of solar ultraviolet light.   At extremely high altitudes there are not enough O_2 molecules for an appreciable amount of O_3 to be formed.   Between 15 and 35 km above the ground, however, the atmosphere is dense enough for the production of O_3 but not so dense that the ozone molecules undergo disruptive collisions too often.   At lower altitudes the ultraviolet light has already been absorbed, so no ozone can come into being there except as a result of lightning strokes.

(*c*)   Ozone is an excellent absorber of ultraviolet light of frequencies not absorbed by other atmospheric constituents, and thus prevents this light from reaching the earth's surface where it would harm most living things.

2.6.   Sketch a mercury barometer and an aneroid barometer and explain their operation.

A mercury barometer (Fig. 2-5*a*) consists of a glass tube closed at one end that is filled with mercury and inverted in a container of mercury in such a way that no air is present in the upper part of the tube.   At the bottom of the tube, the weight of the mercury column is balanced by the downward force of the atmosphere on the mercury surface in the container.   The height of the mercury column at any time is proportional to the atmospheric pressure; the higher the column, the greater the pressure.   A height of 76 cm of mercury corresponds to average sea-level atmospheric pressure.

An aneroid barometer (Fig. 2-5*b*) consists of a sealed metal cylinder with flexible ends.   At low atmospheric pressure the ends expand outward, at high pressure they are forced inward.   A scale linked to one end of the cylinder indicates the pressure at any time.

2.7.   What is a millibar?

The millibar (mb) is a unit of pressure widely used in meteorology.   It is equal to 100 newtons/m², where the newton (N) is the SI unit of force.   Since 1 N = 0.225 lb, 1 mb is equivalent to 0.0145 lb/in². The average pressure of the earth's atmosphere at sea level is 1013 mb.

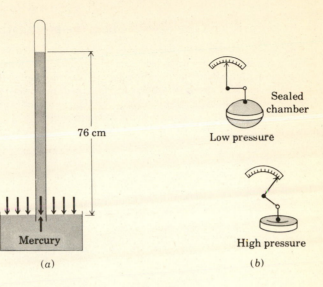

Fig. 2-5

2.8. Light is scattered by small obstacles or irregularities in a medium in its path. The higher the frequency of the light, the more readily scattering occurs. Use this fact to explain why the sky is blue.

When we look at the sky in daytime, what we see is light from the sun that has been scattered by irregularities in the upper atmosphere so that it seems to come from all around the earth. Because blue light has the highest frequency in the visible spectrum, it is scattered to the greatest extent, hence skylight is blue (Fig. 2-6a). At sunrise or sunset, light from the sun must travel a long distance through the atmosphere, and the sun appears red at such times because more blue light is scattered from its direct beam than when it is overhead (Fig. 2-6b).

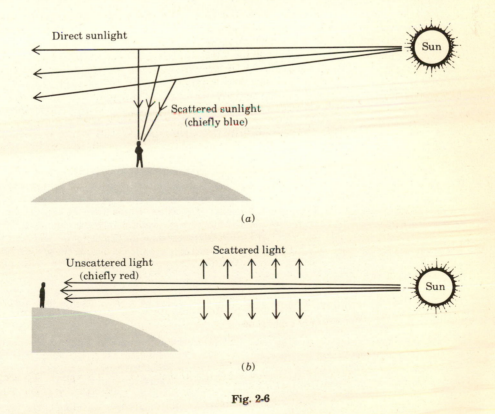

Fig. 2-6

2.9.   **What is the ionosphere and how does it affect the propagation of radio waves?**

The ionosphere is the region of the earth's atmosphere which contains an appreciable number of ions and electrons.   The ionosphere extends through altitudes from about 30 miles to several hundred miles and is produced by the action of solar ultraviolet light and X-rays.   During the day the ionosphere has four layers, $D, E, F_1,$ and F_2 in order of ascending altitude (Fig. 2-7). At night the D layer disappears, the E layer weakens, and the F_1 and F_2 layers coalesce into a single weak F layer.   The D layer partially absorbs radio waves, the other layers reflect them and so make possible long-range radio communication.

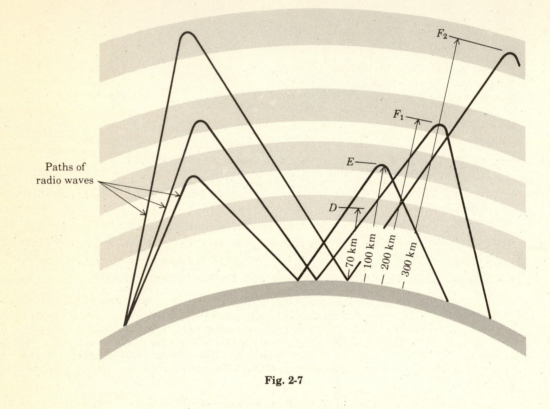

Fig. 2-7

2.10.   **What is a "temperature inversion" in the atmosphere and what is its connection with smog?**

Ordinarily air temperature falls steadily with increasing altitude in the troposphere.   Sometimes, however, a situation arises in which a layer of air aloft is warmer than the air below it; this constitutes a temperature inversion.   (For example, on a clear summer night the earth's surface in a certain region may cool rapidly by radiation, which leads to a layer of cool air near the surface while the overlying air has not changed in temperature by very much.)   Gases emitted by chimneys and vehicle exhausts cannot rise past a temperature inversion because when they reach it their density is greater than the density of the warm air layer.   Hence the inversion acts to trap such gases, whose increased concentration is evident as smog.

2.11.   **What is the "greenhouse effect" and how is it related to the absorption of solar energy by the earth's atmosphere?**

The interior of a greenhouse is warmer than the outside air because sunlight can enter through its windows but the infrared radiation that the warm interior gives off cannot escape through them.   The carbon dioxide and water vapor contents of the atmosphere act as a one-way mirror of this kind of earth as a whole.   The atmosphere is transparent to visible light, which is absorbed by the earth's surface.   The temperature of the surface is thereby increased, which in turn increases the rate at which it emits infrared radiation.   The carbon dioxide and water vapor in the atmosphere absorb the infrared radiation, which leads to a warming of the lower atmosphere.

2.12. Why do the tropics receive more solar energy per unit area than the polar regions?

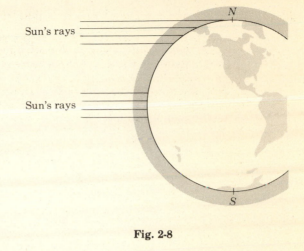

Fig. 2-8

    As shown in Fig. 2-8, sunlight falls almost vertically on the earth's surface near the equator, whereas near the poles it falls at a glancing angle. Thus the sunlight arriving in the tropics is more concentrated than that reaching the polar regions.

2.13. Why do temperatures in the stratosphere increase with altitude?

    The stratosphere is heated largely through the absorption by ozone molecules of ultraviolet radiation from the sun. Because it is heated from above, the temperature of the stratosphere increases with altitude.

2.14. Why do temperatures in the troposphere decrease with altitude?

    The troposphere is heated largely through the absorption by carbon dioxide and water molecules of infrared radiation emitted by the earth's surface. Because it is heated from below, the temperature of the troposphere decreases with altitude.

2.15. The rate at which tropospheric temperatures decrease with altitude is called the *lapse rate* and averages 6.5 °C/km (3.5 °F/1000 ft). What is the *adiabatic lapse rate* and what is its significance for the stability of the troposphere in a particular region?

    In general, an *adiabatic process* is one in which no heat is exchanged between a certain sample of matter and its surroundings. A temperature change in an adiabatic process in a gas can come about only through a change in pressure: compressing a parcel of gas results in a temperature rise, and expanding it results in a temperature drop (see Problem 1.14). Because atmospheric pressure decreases with altitude, a parcel of air that rises through the troposphere will cool as it does so. In the ideal case of dry air and no exchange of heat between the parcel and the air around it, the rate of temperature decrease is called the *dry adiabatic lapse rate* and is about 100 °C/km (5.5 °F/1000 ft).

    If the air contains water vapor, some of the vapor will condense into liquid water as the parcel rises and cools. Because heat is given off when a gas or vapor becomes a liquid, the presence of moisture reduces the rate of cooling with altitude. The *wet adiabatic lapse rate* applies to moist air and varies with temperature (a typical value is 5 °C/km) but is always less than the dry rate.

    A parcel of air in a certain region of the atmosphere will sink if its density is greater than that of the surrounding air and will rise if its density is less. The region is said to be *stable* when there is no tendency for a parcel of air to either sink or rise. This occurs when the actual lapse rate is less than the adiabatic rate. If a parcel of air under these circumstances is displaced upward, it will cool at the adiabatic rate and become colder than the surrounding air, hence its density will be greater than that of its environment and it will sink back to its original height. If the parcel is displaced downward, it will become warmer than the surrounding air, hence its density will be less than that of its environment and it will rise to its original height.

    On the other hand, a region in which the actual lapse rate is greater than the adiabatic rate is *unstable*, because a vertical displacement of a parcel of air will be followed by the continued motion of that parcel. In such a case vertical motion will persist until an altitude is reached at which the lapse rate is less than that of the adiabatic rate. Because the dry and wet adiabatic rates are different, the stability of the atmosphere in a certain region depends upon the relative humidity of the air there if the lapse rate lies between the dry and wet values.

2.16. A parcel of dry air initially at sea level and a temperature of 20 °C moves upward adiabatically to an altitude of 5000 m. What is its temperature at this altitude?

    The dry adiabatic lapse rate is 10 °C/km. Since 5000 m = 5 km, the temperature drop is 50 °C and the final temperature is 20 °C − 50 °C = −30 °C.

2.17. **What are the various ways in which clouds can be formed in an air mass that was originally warm and moist?**

For clouds to form in such an air mass, it must move upward, expand due to the lower pressure, and thereby become cool enough for some of its warm vapor to condense. There are three processes which can accomplish the uplift of an air mass:

1. **Convection.** The warm air rises due to its buoyancy. Cumulus clouds are formed in this way.

2. **Synoptic cooling.** A warm air mass moving horizontally encounters a cooler mass and, being less dense, is forced upward on top of it. Stratus clouds are formed in this way.

3. **Orographic cooling.** A warm air mass moving horizontally encounters a land barrier such as a mountain range and rises. Both cumulus and stratus clouds can be formed in this way.

2.18. **How are clouds classified?**

The three basic types of clouds are *cirrus* (wispy or featherlike), *stratus* (layered), and *cumulus* (puffy or heaped up). A cloud that combines the characteristics of two of these types is designated accordingly, for instance, cirrostratus. A cloud that occurs at a higher altitude than is normal for its type is given the prefix *alto,* as in altostratus. Clouds from which precipitation occurs have the word *nimbus* (Latin for rain) in their names, for instance, nimbostratus.

2.19. **Clouds are divided into four families depending upon the altitudes at which they normally occur, as follows: high (> 7 km), middle (2 to 7 km), low (< 2 km), and clouds with vertical development (base usually < 2 km, top may be > 7 km). Give examples of clouds in each family.**

HIGH: cirrus, cirrostratus, cirrocumulus

MIDDLE: altocumulus, altostratus

LOW: stratus, stratocumulus, nimbostratus

CLOUDS WITH VERTICAL DEVELOPMENT: cumulus, cumulonimbus

2.20. **Why do the water droplets in a cloud not fall to the ground immediately?**

Any object moving through the atmosphere is acted upon by a frictional drag force that increases with its velocity. An object falls faster and faster until the drag force becomes equal to the downward force on it, which is equal to its weight minus the buoyant force. After this balance occurs the object continues to fall at a constant *terminal velocity* whose value depends upon its size and weight. For objects of the same kind and shape, the smaller the size, the smaller the terminal velocity. A typical cloud droplet of 0.02 mm diameter has a terminal velocity of only about 1 cm/s, which means that very little updraft in a cloud is needed to maintain such droplets suspended indefinitely. On the other hand, a typical raindrop of 1 mm diameter has a terminal velocity of 4 m/s, which is 400 times greater, and consequently has a strong tendency to leave the cloud in which it formed.

2.21. **What processes lead to the precipitation of rain or snow from a cloud?**

Precipitation occurs when the ice crystals or water droplets in a cloud grow until they are too heavy to stay aloft and then fall to the ground. Two basic mechanisms are believed to be responsible for precipitation:

1. **Ice crystal growth.** Below 0 °C, ice has a stronger attraction for water molecules than does liquid water. As a result, in a cloud that contains both ice crystals and water droplets below 0 °C, the ice crystals tend to grow at the expense of the water droplets. When such ice crystals become sufficiently heavy, they fall from the cloud and either reach the ground as snow or melt on the way into raindrops.

2. **Coalescence.** Raindrops can form in a cloud that contains water droplets of varying sizes. Because large and small droplets follow different paths at different speeds as the air in the cloud eddies about, collisions between them occur more often than between two droplets of the same size. There is a certain likelihood that colliding drops will stick together, so the result is that the larger drops in the cloud grow at the expense of the smaller ones until they either fall as rain or break up into fragments which in turn begin to grow once more.

In the middle and high latitudes, most precipitation is initiated by ice crystal growth; often the crystals melt into large droplets while still in their parent cloud and grow further by coalescence. In the tropics, coalescence is naturally more significant in low clouds.

2.22. **How do fogs originate?**

Nearly all fogs are caused by the cooling of moist air near the earth's surface. One way in which this can happen is through the loss of heat from the ground by radiation on a clear, still night. If the overlying air is sufficiently humid, its cooling by contact with the ground will lead to saturation and the formation of a *radiation fog*. A radiation fog usually dissipates a few hours after sunrise.

Also common are *advection fogs* produced when warm, humid air flows over colder areas. (*Advection* refers to the horizontal movement of air.) Fogs at sea are normally advection fogs produced by the motion of air from warm parts of the ocean to adjacent cold ones; the notorious fogs of the Grand Banks off Newfoundland occur where the warm Gulf Stream flows past the frigid Labrador Current.

A third type of fog can occur when moist air flows across a region that gradually slopes upward. In this situation the air is cooled adiabatically and a fog may result. Moist air from the Gulf of Mexico is responsible for upslope fogs in the central United States.

2.23. **Clouds are sometimes "seeded" with silver iodide to induce precipitation. Why silver iodide?**

The crystal structure of silver iodide resembles that of ice, hence water molecules in a cloud can readily attach themselves to a silver iodide crystal. Silver iodide crystals are thus efficient condensation nuclei and so promote precipitation from a cloud.

Supplementary Problems

2.24.   Why does the earth have an atmosphere whereas the moon does not?

2.25.   Distinguish between the troposphere and the tropopause.

2.26.   What characterizes (*a*) the tropopause, (*b*) the stratopause, and (*c*) the mesopause?

2.27.   What is the color of space to an astronaut above the earth's surface?

2.28.   What is insolation?

2.29.   The graph in Fig. 2-9 shows variations of atmospheric temperature with altitude that correspond to the following lapse rates: average, dry adiabatic, wet adiabatic, and inversion. Label the curves accordingly.

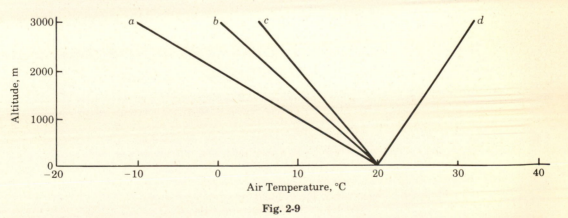

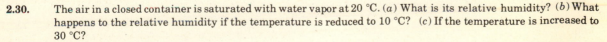

Fig. 2-9

2.30.   The air in a closed container is saturated with water vapor at 20 °C. (*a*) What is its relative humidity? (*b*) What happens to the relative humidity if the temperature is reduced to 10 °C? (*c*) If the temperature is increased to 30 °C?

2.31.   Why does the air in a heated room tend to be dry?

2.32. Why does dew form during clear, calm summer nights?

2.33. What does "dew-point temperature" mean?

2.34. How are clouds and fog related?

2.35. What do high-altitude clouds consist of? Low-altitude clouds?

2.36. What is the significance for weather phenomena of dust and salt particles in the atmosphere?

2.37. What is supercooled water?

2.38. How do clouds that consist of water droplets differ in appearance from those that consist of ice crystals?

2.39. The presence of a tropical island beyond the horizon is usually announced to sailors and fliers by a stationary bank of cumulus clouds. How do such clouds form?

Answers to Supplementary Problems

2.24. The gravitational field of the moon is not sufficient to prevent the escape of the rapidly moving gas molecules in an atmosphere. The gravitational field of the more massive earth is great enough to retain an atmosphere.

2.25. The troposphere is the dense lower part of the atmosphere in which clouds, storms, and other weather phenomena occur. The tropopause is the upper limit of the troposphere and divides it from the clear, cold air of the stratosphere above.

2.26. (a) A temperature minimum; (b) a temperature maximum; (c) a temperature minimum.

2.27. Since there is no atmosphere in space, there is no scattered sunlight, and space appears black.

2.28. "Insolation" stands for *in*coming *sol*ar radi*ation* and refers to the solar energy arriving at the top of the earth's atmosphere.

2.29. a = dry adiabatic; b = average; c = wet adiabatic; d = inversion.

2.30. (a) 100%; (b) the air remains saturated and so the relative humidity remains 100%, while the excess water vapor condenses out; (c) the relative humidity decreases.

2.31. The outside air has a low moisture content because it is cold, even though its relative humidity may be high. When this air is heated, its moisture content remains the same, hence its relative humidity decreases.

2.32. On such a night the earth's surface cools by radiation. The air in contact with the surface cools also until it becomes saturated with water vapor, which then condenses into droplets of liquid water.

2.33. The dew-point temperature is the lowest temperature to which a sample of air can be cooled without the condensation of water vapor. The lower the dew point, the smaller the concentration of water vapor in the air sample.

2.34. A fog is a cloud at ground level.

2.35. Ice crystals; water droplets.

2.36. Dust and salt particles provide the nuclei around which water vapor condenses to form clouds, fog, rain, and snow.

2.37. Although the normal freezing point of water is 0 °C, water droplets in the atmosphere sometimes remain liquid at temperatures down to about −40 °C. It is thought that the absence of particles that can act as crystallization nuclei enables such *supercooled water* to exist.

2.38. Water clouds usually have sharp boundaries. Ice clouds have a fibrous appearance with irregular boundaries.

2.39. During the day the sun warms an island, and the air over it is heated in turn and rises. The replacement air that flows toward the island is moist, and as it rises it cools by expansion, with the excess water vapor condensing into cumulus clouds.

Chapter 3

Weather

WINDS

Winds are horizontal movements of air that take place in response to pressure differences in the atmosphere. The greater the difference between the pressures in two regions, the faster the air between them will move. All pressure differences between places on the earth's surface can be traced, directly or indirectly, to temperature differences. If one region is warmer than its surroundings, for example, the air above it is heated and expands. The hot air rises, leaving behind a low-pressure zone into which cool air from the high-pressure neighborhood flows. The horizontal flow toward the heated region at low altitudes is balanced by a horizontal flow outward of air that has risen, which cools and sinks to replace the air that has moved inward. In this way *convection cells* come into being that convert temperature differences into pressure differences and thus cause winds to occur. On a large scale, the differences between the solar heating of the equatorial and polar regions are what power the general circulation of the lower atmosphere.

CORIOLIS EFFECT

The rotation of the earth influences the path of an object that moves above its surface as this path is seen by an observer on the surface. In the northern hemisphere a path that would be a straight line across a stationary earth appears instead to be curved to the right (as the object recedes from the observer); in the southern hemisphere the curvature is to the left. Only motion along the equator is not affected. This phenomenon is called the *Coriolis effect*.

The Coriolis effect is responsible for converting the north-south convection currents that would be caused on a nonrotating earth by the uneven distribution of solar heating into winds that have easterly (that is, out of the east) and westerly (out of the west) components (see Fig. 3-1).

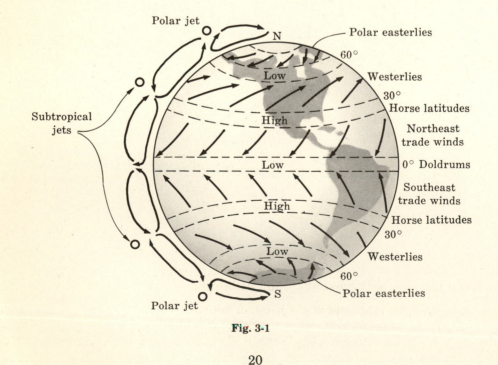

Fig. 3-1

GENERAL CIRCULATION OF THE ATMOSPHERE

When the winds of the world are averaged over a long period of time, transient fluctuations disappear to leave a large-scale *general circulation* of the lower atmosphere. The principal features of the general circulation are shown in Fig. 3-1. Conspicuous are the easterly winds of the polar regions, the *prevailing westerlies* of the middle latitudes, and the northeast and southeast *trade winds* of the tropics. An equatorial belt of low pressure, the *doldrums,* separates the trade winds of each hemisphere, and belts of high pressure, the *horse latitudes,* separate the trade winds from the prevailing westerlies; winds in these belts are weak and erratic.

With increasing altitude the regions of westerly winds broaden until almost the entire flow of air is west-to-east at the tropopause. The westerly flow aloft is not uniform but contains narrow cores of high-velocity winds called *jet streams*. The jet streams form wavelike zigzag patterns around the earth that change continuously and give rise to the variable weather of the middle latitudes by their effect on air masses closer to the surface.

CYCLONES AND ANTICYCLONES

Winds in the middle latitudes are associated with weather systems called *cyclones* and *anticyclones* that are several hundred to a thousand or more miles across and move from west to east. At the center of a cyclone the air pressure is low, and as air rushes in toward it the moving air is deflected toward the right in the northern hemisphere and toward the left in the southern, because of the Coriolis effect. As a result cyclonic winds blow in a counterclockwise spiral (as viewed from above) in the northern hemisphere and in a clockwise spiral in the southern hemisphere (Fig. 3-2). An anticyclone is centered on a high pressure region from which air moves outward. The Coriolis effect therefore causes anticyclonic winds to blow in a clockwise spiral in the northern hemisphere and in a counterclockwise spiral in the southern hemisphere. These spirals are conspicuous in cloud formations photographed from earth satellites.

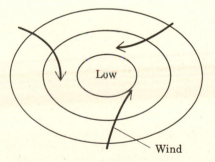
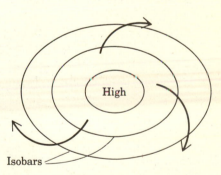

Cyclone in the Northern Hemisphere   Anticyclone in the Northern Hemisphere

Fig. 3-2

As a rule, cyclones bring unstable weather conditions with clouds, rain, strong winds, and abrupt temperature changes. The weather associated with anticyclones, on the other hand, is usually settled and pleasant with clear skies and little wind.

Middle-latitude cyclones originate at the *front,* or boundary, between the cold polar air mass and the warmer air mass adjacent to it. It is common for a kink to develop in this front with a wedge of warm air protruding into the cold air mass. This produces a low-pressure region which moves eastward as a cyclone. The eastern side of the warm air wedge is a *warm front* since warm air moves in to replace cold air there: the western side is a *cold front* since cold air replaces warm air.   After a few days the cold front, which moves faster, overtakes the warm front to force the wedge of warm air upwards, and soon afterward both it and the cyclone disappear.

CLIMATE

The *climate* of a region refers both to its average weather over a period of years and to the typical amounts by which the various weather elements (notably temperature and precipitation) vary during each year.   The Köppen system of climate classification has five principal categories:

A. **Tropical rainy climates.**   The average monthly temperature never goes below 18 °C (64.4 °F) and there is little seasonal variation.   Annual rainfall exceeds the water lost by evaporation.   Southern Florida and the lowlands of Mexico and Central America have such a climate.

B. **Dry climates.**   The water lost by evaporation exceeds that brought by precipitation. The southwestern part of the United States has such a climate.

C. **Warm temperate rainy climates.**   There are distinct summer and winter seasons, with the average temperature of the coldest month lower than 18 °C (64.4 °F) but higher than −3 °C (26.6 °F). The United States east of the Rockies and south of a line between New York and Denver has such a climate.

D. **Cool snow-forest climates.**   The average temperature of the coldest month is lower than −3 °C (26.6 °F) and that of the warmest month is higher than 10 °C (50 °F).   The New England and North Central states together with most of Canada have such a climate.

E. **Polar climates.**   The average temperature of the warmest month remains below 10 °C (50 °F).   Trees do not thrive in such climates, which exist in northern Canada and Alaska.

Solved Problems

3.1.   The earth is closest to the sun in January, but January is a winter month in the northern hemisphere.   Why?

The earth's axis of rotation is tilted with respect to its axis of revolution around the sun (Fig. 3-3).   As a result the daylight side of the northern hemisphere is tilted away from the sun in January, which means that sunlight strikes the northern hemisphere at a glancing angle and delivers less energy per square meter of surface than during the summer when the northern hemisphere is tilted toward the sun.   The effect of the tilted axis is more than enough to counterbalance the relative closeness of the sun during the northern winter.

3.2.   In the northern hemisphere, the longest and shortest days occur respectively in June and December, as shown in Fig. 3-3, but the warmest and coldest weather of the year occur respectively a month or two later.   What is the reason for these time lags?

The heat capacities of land, sea, and air are very large, and when the rate of arrival of solar energy changes considerable energy must be absorbed or lost by a region before it reaches equilibrium.   Since the difference between the rates of energy absorption and energy loss is always small, the temperature of the surface cannot change rapidly enough to keep pace with changes in the rate at which solar energy arrives, hence the time lags in seasonal weather conditions.

3.3.   Distinguish between an isobar and a millibar.

An isobar is a line on a weather map that joins points which have the same atmospheric pressure.   A millibar (mb) is a unit of pressure; average sea-level atmospheric pressure is 1013 mb.

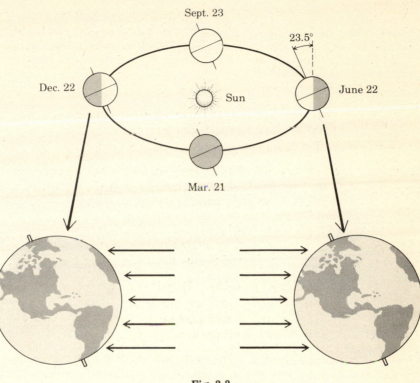

Fig. 3-3

3.4. Why does a rising body of air become cooler?

   Atmospheric pressure decreases with altitude, hence a rising body of air expands. An expanding gas does work on its surroundings, and its loss of energy means a drop in temperature. Dry rising air cools at the rate of about 1 °C per 100 m.

3.5. Use a diagram to show the convection currents around a heated area on the earth's surface.

   See Fig. 3-4.

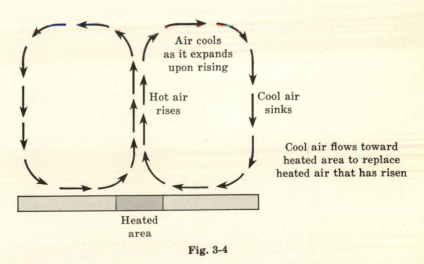

Air cools
as it expands
upon rising

Hot air
rises

Cool air
sinks

Cool air flows toward
heated area to replace
heated air that has risen

Heated
area

Fig. 3-4

3.6. (a) On summer days coastal regions often experience an onshore wind. What is the origin of such a sea breeze? (b) At night the sea breeze stops and is often replaced by an offshore wind. What is the origin of such a land breeze?

(a) Sunlight causes the land to warm up fairly rapidly in the morning, since it is absorbed in a thin surface layer. The water temperature changes very little, partly because the incoming solar energy is shared by a thicker layer of water and partly because the specific heat capacity of water is large. The air over the warm land becomes warm in turn and rises by convection, whereupon cooler, denser air from the sea—the sea breeze—sweeps in to replace it.

(b) At night the land cools rapidly by radiation while the sea surface remains at about the same temperature as during the day because heat transfer is more efficient in water than in rock and soil. When land and sea are at the same temperature, the sea breeze stops. If the land cools still further, air warmed by the sea rises and cool air sweeps off the land to replace it.

3.7. In summer, the southwest monsoon of warm, moist air flows northeastward over India and Southeast Asia to bring heavy rain. In winter, the direction of flow reverses and the northeast monsoon of cold, dry air from Siberia invades this region. How do these monsoons originate?

Continental temperatures change with the seasons to a greater extent than oceanic temperatures. In summer, when a continent is warmer than the ocean bordering it, the flow of air is toward the land, where it rises; in winter, the cool continental air sinks and the flow is away from the land. Thus the monsoons are large-scale analogs of the sea and land breezes of Problem 3.6.

3.8. What is a *katabatic wind*?

A katabatic wind consists of cold air that flows down a mountainside because its density is greater than that of the air over the adjacent valley. Such winds occur at night in mountainous regions because the air in contact with mountains cools more rapidly than air at the same altitude over valleys. On a large scale, katabatic winds arise when air is cooled as it moves across an elevated region (especially one containing glaciers and snow fields) and then descends along the slopes of the region. The *bora* of the Adriatic Sea and the *mistral* of southern France are examples of such winds.

3.9. The northeast and southeast trade winds meet in a belt called the *doldrums*. What is the characteristic weather of the doldrums?

The doldrums are at the equator, so it is quite warm there with considerable evaporation of water and thus high humidity. The air flow is largely upward, so surface winds are light and erratic. The rising currents of moist air lead to considerable rainfall.

3.10. Why are most of the world's deserts found in the *horse latitudes,* which separate the trade winds from the prevailing westerlies in both hemispheres?

The air flow in the horse latitudes is largely downward. Descending air is warmed by compression and its relative humidity decreases, hence there is little rainfall in the horse latitudes.

3.11. (a) When you face a wind associated with a cyclone in the northern hemisphere, in what approximate direction will the center of low pressure be? (b) In what direction will the center of low pressure be if you do this in the southern hemisphere?

(a) The circulation of air around a low-pressure region in the northern hemisphere is counterclockwise since air flowing toward the region is deflected to its own right by the Coriolis effect. Hence when you face the wind, the center of low pressure will be on your right.

(b) In the southern hemisphere the Coriolis effect causes moving air to be deflected to its own left, so the circulation around a low-pressure region is clockwise there. When you face the wind in this case, the center of low pressure will be on your left.

3.12. Why is the sky in the vicinity of a cyclone cloudy whereas it is clear in the vicinity of an anticyclone?

A cyclone is a region of low pressure, and air flowing into it rises in an upward spiral. The rising air cools and its moisture content condenses into clouds. An anticyclone is a region of high pressure, and air flows out of it in a downward spiral. The descent warms the air and its relative humidity accordingly drops, hence condensation does not occur.

3.13. Sketch a typical mature cyclone as it might appear on a weather map of the northern hemisphere and identify its main features.

See Fig. 3-5.

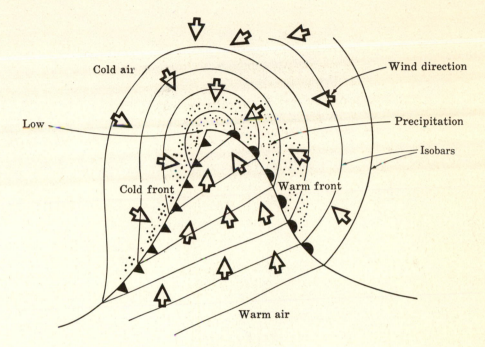

Fig. 3-5

3.14. Draw cross-section diagrams of a typical warm front and cold front. Indicate the types of clouds that might be expected in each case.

See Fig. 3-6.

3.15. What is an *occluded front*? At what stage in the evolution of a cyclone does it occur?

In a typical cyclone, a wedge of eastward-moving warm air penetrates a cold air mass. There is a warm front on the east and a cold front on the west of the wedge. The cold front moves faster than the warm front, and when it overtakes the warm front their intersection is lifted above the ground as the cold air burrows under the warm air of the wedge. The formation of such an occluded front is the last stage in the evolution of a cyclone, which soon afterward disappears.

3.16. Illustrate the life cycle of a middle-latitude cyclone in the northern hemisphere and describe the behavior of the various fronts.

See Fig. 3-7. (*a*) The polar front between cold polar air flowing westward and warm middle-latitude air flowing westward is straight. (*b*) A kink develops in the polar front with a warm front to the east and a cold front to the west. (*c*) The kink becomes a mature cyclone with its characteristic counterclockwise circulation and well-developed warm and cold fronts. (*d*) The cold front has begun to overtake the warm front to form an occluded front. (*e*) All the warm air in the kink has been replaced by cold air at the surface in the final stage of decay. The whirl of cold air will soon disappear.

3.17. What are the characteristic properties of the air in the following air masses: (*a*) continental polar; (*b*) continental tropical; (*c*) maritime polar; (*d*) maritime tropical?

(*a*) Cold, dry, stable. (*b*) Hot, dry, unstable. (*c*) Cool, moist, unstable in winter; cool, dry, stable in summer. (*d*) Warm, moist, unstable.

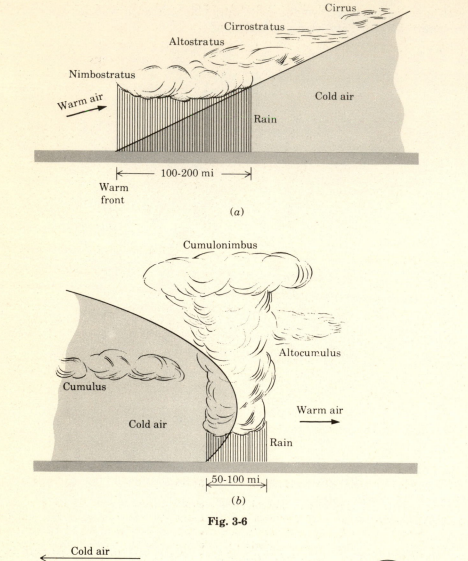

Fig. 3-6

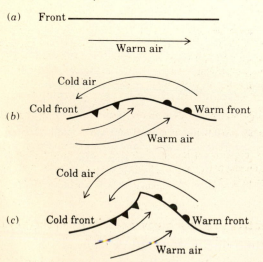

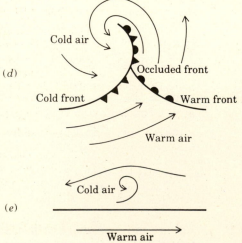

Fig. 3-7

3.18. Sketch the air masses that affect the weather of North America. Are all of them present at all seasons of the year?

See Fig. 3-8. The continental tropical air mass is absent in the winter.

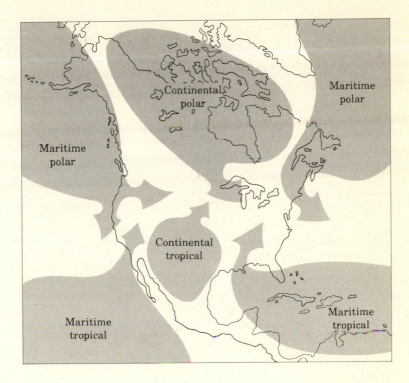

Fig. 3-8

3.19. Distinguish between a hurricane and a tornado.

A hurricane is a large, violent tropical storm typically a hundred miles in diameter whose winds spiral inward and upward at velocities of 75 mi/hr or more around an "eye" of low pressure. Heavy rainfall accompanies the passage of a hurricane except in the eye, which may be 10 or 20 miles across. Most hurricanes occur on the western sides of the Pacific, Indian, and North Atlantic oceans during the late summer and early fall and their most violent phases last for a few days to a week or so. Hurricanes usually move at 10 to 30 mi/hr but may move faster or remain in one place for a day or more.

A tornado is a small, funnel-shaped rotational storm several hundred feet in diameter that appears to descend from a cumulonimbus (thunderstorm) cloud. Velocities in a tornado are apparently several hundred mi/hr, though the tornado itself moves at only 20 to 40 mi/hr; a tornado usually last for less than an hour. Tornados most often occur on central continental plains, notably those of the United States and Australia, but are also found at sea, where they are called *waterspouts* and are less violent.

3.20. The energy released by a hurricane per day exceeds the energy consumed by mankind throughout the world per year. Where does all this energy come from?

A hurricane draws in vast quantities of warm, moisture-laden air from distances of hundreds of miles around it. The air rises near the center of the hurricane and heavy precipitation occurs, as much as a billion tons of water per hour. The hurricane derives its energy from the heat of vaporization liberated when the water vapor condenses into water droplets.

3.21. What is the direction of air flow in the central eye of a hurricane? Why is the sky clear over the eye?

The air flow is downward and, near the bottom, outward. As the air descends, it is compressed and therefore warmed, which decreases its relative humidity below saturation. Clouds come into being only in saturated air.

3.22.   Average annual rainfall varies with latitude as shown in Fig. 3-9.   Why does the curve have this particular shape?

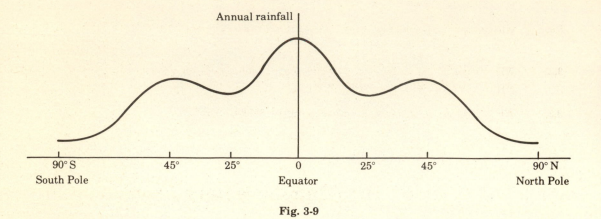

Fig. 3-9

The trade winds in both hemispheres bring warm, moist air to the equator where it rises and cools, which causes the moisture to precipitate as rain. The rainfall minimum at about 25° on either side of the equator occurs at the boundary (the horse latitudes) between the trade winds and the westerlies where the air flow is largely downward.   As air descends, it becomes warmer and its relative humidity decreases, hence there is little tendency for rain to fall.   Poleward of the horse latitudes rainfall increases because of the frequent cyclones that form at the polar front; the maximum occurs at about 45°.   Precipitation is sparse in the polar regions because temperatures there are too low to permit the accumulation of much moisture in the air.

3.23.   What is a *degree day*?

A degree day at a certain place corresponds to a day in which the average temperature differs from an arbitrary reference temperature (65 °F is common) by 1 °F.   An average temperature of 50 °F on a certain day thus corresponds to 15 degree days.   The total number of degree days over a period of time is a measure of the amount of domestic heating or cooling required to maintain a comfortable interior temperature during that period.   Annual heating degree days based on 65 °F in the United States varies from a few hundred in southern Florida to 9000 or more in northern and Rocky Mountain states.

Supplementary Problems

3.24.   What is the source of energy that is manifested in weather phenomena?

3.25.   What are the two mechanisms by which energy of solar origin is transported around the earth?   Which is the most important?

3.26.   What is the direction of the prevailing winds of the middle latitudes in each hemisphere?

3.27.   Where in the atmosphere do the jet streams occur?   What is their general direction?

3.28.   A yachtsman is planning to sail from the U.S. to England and later to return home.   What routes should he follow across the Atlantic in order to have his course downwind as much of the time as possible?

3.29.   The prevailing westerlies of the middle latitudes of the southern hemisphere are considerably stronger than those of the northern hemisphere.   Why?

3.30.   How does the weather associated with a typical cyclone differ from that associated with a typical anticyclone?

3.31.   What is the approximate sequence of wind directions when the center of a cyclone passes north of an observer in the northern hemisphere?

3.32.   What is the approximate sequence of wind directions when the center of an anticyclone passes south of an observer in the northern hemisphere?

3.33.   What is the difference between the rainfall that accompanies the passage of a warm front and that which accompanies the passage of a cold front?

3.34.   What is the usual lifetime of a cyclone in the middle latitudes?

Answers to Supplementary Problems

3.24.   Solar radiation reaching the earth.

3.25.   Winds and ocean currents carry energy around the earth in the forms of warm air and warm water respectively. Winds are more effective in energy transport than ocean currents.

3.26.   Westerly (from the west) in both hemispheres.

3.27.   The jet streams occur near the top of the troposphere and flow from west to east.

3.28.   To England: northeast and then east in order to take advantage of the prevailing westerlies. Back to the U.S.: first south to the trade winds, then west, and finally north.

3.29.   Large land masses in the middle latitudes of the northern hemisphere obstruct the westerlies. In the southern hemisphere the middle latitudes are almost free of land, which permits high average wind velocities to develop; hence the term "roaring forties" for the latitude band 40 °S–50 °S.

3.30.   Anticyclonic weather is generally steady with a relatively constant temperature, clear skies, and light winds. Cyclonic weather is unsettled with rapid changes in temperature that accompany the passages of cold and warm fronts, cloudy skies, rain, and fairly strong, shifting winds.

3.31.   From the southwest—from the west—from the northwest.

3.32.   From the northwest—from the west—from the southwest.

3.33.   Rainfall associated with the passage of a warm front is generally lighter and of longer duration than that associated with the passage of a cold front.

3.34.   Three to five days.

Chapter 4

The Oceans

OCEAN WATER

During the early history of the earth the gases in its initial atmosphere escaped into space. The constituents of the present atmosphere plus the water of the hydrosphere are believed to have emerged over a long period of time as a by-product of volcanic activity from the rocks of the earth's interior, where they were incorporated in various minerals.

The oceans cover about 71% of the earth's surface and contain about 97% of the water in the hydrosphere. The average salinity of ocean water is 3.5%; most of the ions are Na^+ and Cl^-, though many other ions are also present. Ocean water near the surface is saturated with atmospheric gases, whose concentrations decrease with depth.

The water in the top hundred meters of the ocean varies in temperature with location and season, and may be 20 °C or more. In the next km the temperature drops to a few °C and remains just above the freezing point to the ocean bottom, whose depth averages 3.7 km.

WAVES

Wind blowing across the surface of a body of water produces waves. The wave height depends upon the amount of energy transferred to the water and therefore increases with wind velocity, the time during which the wind has blown from the same direction, and the distance (called *fetch*) over which the wind has blown. When a wave passes a certain place in the ocean, water particles at and near the surface move in circular orbits (Fig. 4-1). At the wave crest, the surface water moves in the direction of the wave; at the wave trough, it moves in the opposite direction. The orbital motion becomes negligible at a depth of about $\lambda/2$, where λ is the wavelength.

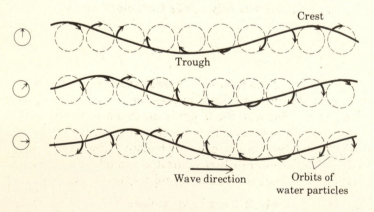

Fig. 4-1

CURRENTS

The progress of a wave involves the transport of energy but not of water. However, a wind whose direction remains more or less constant will produce a net motion of surface water called a *current*. The wind patterns of the general circulation of the atmosphere lead to corresponding patterns of ocean currents, with the Coriolis effect playing a part. The latter patterns take the form of huge whirlpools, or *gyres,* in each ocean basin: gyres of clockwise flow in the North Atlantic and North Pacific Oceans, gyres of counterclockwise flow in the South Atlantic, South Pacific, and Indian Oceans (Fig. 4-2). In addition, smaller counterclockwise gyres are present in the extreme northern parts of the North Atlantic and North Pacific. Eastward *equatorial countercurrents* flow in the doldrums of the Atlantic and Pacific between the gyres to the north and south, and a continuous eastward current flows around the entire Southern Ocean.

Fig. 4-2

The westward-moving currents in the major gyres, which are propelled by the trade winds, are warmed by the tropical sun, and much of the added heat is retained in their further progress poleward and then eastward. Thus the North Equatorial Current is quite warm when it turns north as the Gulf Stream, and as the North Atlantic Drift it significantly moderates the climate of northern Europe by heating the westerly winds that blow across it.

THE TIDES

The twice-daily rise and fall of the tides can be traced to the gravitational forces exerted by the moon and, to a lesser extent, by the sun. Each of these bodies attracts different parts of the earth with slightly different forces. Because the moon is much closer to the earth than the sun is, the variations in the forces it exerts are greater than those in the forces exerted by the sun, even though the forces themselves are stronger in the case of the sun. For this reason the moon is chiefly responsible for the tides, with the solar influence limited to modifying the tidal range depending on the position of the sun with respect to the earth and moon.

The tidal bulges A and B in Fig. 4-3 are held in place by the moon as the earth rotates under them, thus causing the periodicity of the tides. Water at A is closest to the moon, and is heaped up by its gravitational pull. Water at B is farthest from the moon, and is least attracted by it; this water accordingly tends to be left behind, so to speak, as the earth is pulled away from under it due to the revolu-

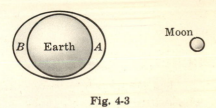

Fig. 4-3

tion of the earth and moon about their common center of mass. (The earth's pull on the moon, which causes it to move in an orbit, has a counterpart in the moon's pull on the earth, which also moves in an orbit, although a much smaller one. Earth and moon may be thought of as the two ends of a dumbbell that is rotating about its balance point, which happens to lie within the earth about 4700 km from the center.   It is this center of mass that moves in an elliptical orbit around the sun.)

Solved Problems

4.1.   List the chief reservoirs of the earth's water content in the order of the amount of water each contains.

      1.   Seas and oceans

      2.   Icecaps and glaciers

      3.   Groundwater

      4.   Lakes and rivers

      5.   Atmospheric moisture

4.2.   Sketch the profile of a typical continental margin and identify its main features.

See Fig. 4-4.

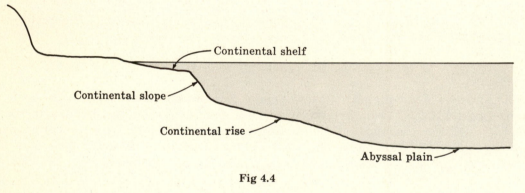

Fig 4.4

4.3.   What are plankton, where do they live, and why are they important to life on land as well as to aquatic life?

Plankton are minute oceanic plants (phytoplankton) and animals (zooplankton).   Most plankton are found in the upper 10 m or so of the oceans where sunlight penetrates, and constitute the first step in the food chain of all other forms of aquatic life such as fish, shellfish, and mammals (whales, porpoises).   The mass of phytoplankton in the ocean exceeds the mass of land vegetation, and phytoplankton are thus important in maintaining the global oxygen-carbon dioxide balance.

4.4.   Why is marine life most abundant in the colder parts of the oceans?

Cold ocean water contains more of the dissolved CO_2 and O_2 required by marine plants and animals because gases are more soluble in cold water than in warm water.   Another reason is that vertical mixing of ocean water is impeded by the presence of a layer of warm surface water, so organic debris that sinks to the bottom in warm parts of the ocean is never returned to the surface where it can provide nutrients (especially nitrates and phosphates) to living things. In cold parts of the ocean, on the other hand, the upwelling of bottom water helps to feed plants and animals near the surface where sunlight and dissolved gases are also available.

4.5.   The salinity of the Red Sea is about four times greater than that of the Baltic Sea.   Why?

The Red Sea lies in the tropics, where extensive evaporation increases the salt concentration.   Also, the adjoining lands are arid and do not contribute much fresh water to dilute the sea water.   The Baltic Sea is near the Arctic, where average temperatures are low and hence evaporation is minor; furthermore, the melting of winter snow contributes considerable fresh water to the Baltic in the summer months.

4.6.   Although the salinity of sea water varies with location, the relative proportions of the various ions in solution are almost exactly the same everywhere regardless of local circumstances.   What is the significance of the latter observation?

Because their waters must be thoroughly mixed in the course of time to obtain a uniform relative composition, the seas and oceans of the world cannot be static bodies but must contain large-scale currents, both vertical and horizontal.

4.7.   Why do icebergs float?

Water expands as it freezes, and so has a lower density, because water molecules are farther apart in the regular structure of an ice crystal than they are in the irregular arrangements characteristic of liquid water. Since ice has a lower density than liquid water, it floats

4.8.   Where do icebergs originate?

Nearly all icebergs consist of chunks of ice that have broken away from the continental glaciers of Greenland and Antarctica.   The Greenland icebergs are usually smaller and hence shorter-lived (a year or two at most) than the Antarctic icebergs, which may last for ten years or even more.

4.9.   Even in the intense cold of the polar regions, sea ice is seldom more than 3 m thick. (Icebergs are huge masses of ice that have broken off from icecaps or glaciers that formed from accumulated snow on a land base.)   Why is sea ice so relatively thin?

There are several reasons.   The chief one is that, as the water under a surface layer of ice is cooled, it becomes denser and sinks, to be replaced by warmer water from underneath.   Another reason is that ice itself acts as an insulator.   Also, as seawater freezes, its salt content is not incorporated in the ice crystals but stays behind to increase the salinity of the remaining water, whose freezing point is accordingly lowered.

4.10.   (*a*) Would sea level change if the ice floating in the polar oceans were to melt?   (*b*) If the glaciers of Greenland and Antarctica were to melt?

(*a*)   No, because the proportion of ice above the water surface is exactly equal to the shrinkage that occurs when ice melts.

(*b*)   Yes, because this ice is now on land and if melted would flow into the oceans.

4.11.   The California Current along the coast of California is cooler than the ocean to the west. How does this fact account for the frequency of fogs on this coast?

When warm, moist air from the west blows over the colder California Current, the air's temperature drops and moisture from the now supersaturated air condenses into droplets to form a fog.

4.12.   Waves always approach a sloping beach so that their crests are parallel to the shoreline, no matter what the original direction of the waves is.   Why?

As a wave moves obliquely shoreward, its near-shore end encounters shallow water where friction with the bottom slows it down.   More and more of the wave is slowed as it continues to move, and the slowing becomes more pronounced as the water gets shallower.   As a result the whole wavefront swings around until it is moving almost directly shoreward (Fig. 4-5).   This effect is known as *refraction* and it occurs whenever a wave of any kind moves through a region in which the wave speed changes so that part of each wavefront moves faster or slower than the remainder.

4.13.   Why do waves steepen and eventually "break" when they approach a beach?

In shallow water, friction with the bottom reduces the speeds of water particles in the trough of a wave to a greater extent than those of water particles in the crest.   In consequence each crest begins to overtake the trough ahead of it, which steepens the front of the wave.   When the wave height is about 1.3 times the average water depth the crest overtakes the trough and falls in a turbulent cascade, which is what is meant by saying that the wave breaks.

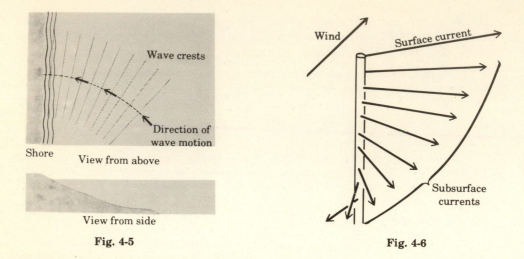

Fig. 4-5   Fig. 4-6

4.14. Would you expect the direction of a wind-driven current to be exactly the same as that of the wind responsible for it?

In open water, the direction of a wind-driven current is different from that of the wind that produced it, as a consequence of the earth's rotation. This phenomenon is an example of the Coriolis effect mentioned in Chapter 3. The deflection of the current relative to the wind is to the right in the northern hemisphere and to the left in the southern hemisphere.

According to theory, a steady wind blowing over an infinitely large and deep ocean will lead to motion of the surface water layer 45° from the wind direction. Successively lower layers are deflected to greater and greater extents, but viscosity (internal friction) causes the velocities of the water in these layers to decrease with depth. At a depth of the order of 100 m in this idealized situation, the direction of water movement is 180° different from the wind direction while its velocity has become negligibly small. If an arrow is used to represent the direction and magnitude of the water velocity in each layer, the result is a pattern called the *Ekman spiral* (Fig. 4-6).

Although the Ekman spiral has been observed in the atmosphere and in laboratory experiments, it has not been clearly demonstrated in actual oceans. Nevertheless there is considerable evidence that the net transport of water due to a prolonged wind is indeed to the right of the wind direction in the northern hemisphere and to the left in the southern, as predicted.

4.15. In a number of regions of the world cold water rises to the surface from considerable ocean depths. What might cause such upwelling to occur? Why does it not take place everywhere in the oceans?

A common cause of upwelling is a steady wind whose effect is to drive warm surface waters away from a coast. The cold bottom water then rises to the surface to replace it. A north wind along the California coast, for example, leads to a westward flow of surface water because of the Coriolis effect (see Problem 4.14), with a consequent upwelling of cold water which brings an increased supply of nutrients and in turn a larger fish population (see Problem 4.4). A south wind along the Peruvian coast has a similar effect, where the deflection is to the left of the wind in the southern hemisphere. A reversal in the wind direction brings warm water to these coasts, with detrimental results to local fisheries. Normally upwelling does not occur because warm water is less dense than cold water and hence floats on top of it.

4.16. (*a*) Under what circumstances is the range of the tides a maximum? (*b*) Under what circumstances is it a minimum? (*c*) What phases of the moon are associated with these extremes?

(*a*) When the sun and moon are in line with each other and with the earth, their tide-producing forces add together and high tides are at their highest, low tides are at their lowest. Such tides are called *spring tides* (Fig. 4-7*a*).

(*b*) When the sun-earth and moon-earth lines are perpendicular, the tide-producing forces of the sun and moon are in opposition, and high tides are at their lowest, low tides are at their highest. Such tides are called *neap tides* (Fig. 4-7*b*).

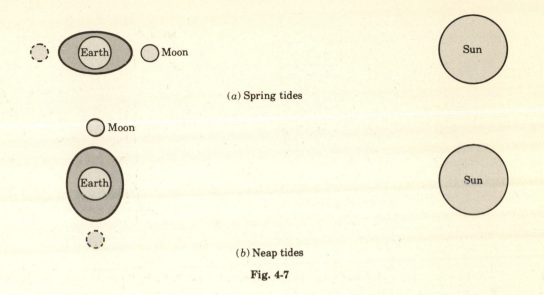

(a) Spring tides

(b) Neap tides

Fig. 4-7

(c) Spring tides occur at new moon and full moon, neap tides occur at first quarter and last quarter (when half the moon's disk is illuminated).

4.17.   The earth makes a complete rotation on its axis once every 24 hr.   Why is the interval between successive high tides 12 hr 25 min instead of 12 hr?

The moon revolves around the earth in the same direction as the earth's rotation and takes 29.5 days to circle the earth relative to the sun; that is, 29.5 days elapse from new moon to new moon.   Thus the moon is directly overhead a certain place on the earth 24 hr/29.5 days = 0.81 hr later each day, which is 50 min.   There are two tides a day, and half of 24 hr 50 min is 12 hr 25 min.

4.18.   How do the tides affect the rotation of the earth?

The moon's attraction holds the tidal bulges in place (Fig. 4-3) as the earth rotates underneath them.   Friction between the solid earth and the seas and oceans acts as a gigantic brake to slow down the earth's rotation, so that the length of the day is gradually increasing.

4.19.   What is a *tsunami*?

A tsunami consists of waves produced in an ocean by an earthquake in the ocean floor.   The waves are very long and low in open water, and move at several hundred miles an hour.   When the waves reach shallow water near a coast, their heights may increase dramatically and they can cause considerable damage on shore.

Supplementary Problems

4.20.   Which ions are predominant in sea water?

4.21.   How does the average depth of the ocean basins below sea level compare with the average height of the continents above sea level?

4.22.   Sunlight cannot penetrate to the bottom waters of the oceans, so plant life is absent there.   Nevertheless these waters contain an abundance of animals, largely invertebrates.   Upon what do these creatures feed?

4.23.   Approximately what proportion of the earth's land area is permanently covered by ice?   Where is the largest deposit of such ice?

4.24.   The prevailing summer winds of the east coast of the United States are southwesterly.   Would such winds tend to produce onshore or offshore surface currents?

4.25. A wind begins to blow over the surface of a calm body of deep water. What factors govern the height of the waves that are produced?

4.26. Why do submarines submerge in storms?

4.27. Do ocean waves actually transport water from one place to another? If not, what if anything do such waves transport?

4.28. (a) If you were planning to drift in a raft across the North Atlantic from the U.S. to Europe by making use of ocean currents, what would your route be? (b) If you were planning to drift from Europe to the U.S., what would your route be?

4.29. England and Labrador are at about the same latitude on either side of the North Atlantic Ocean, but England is considerably warmer than Labrador on the average. Why?

4.30. The average annual range of temperatures in the middle latitudes of the northern hemisphere is considerably greater than that in the corresponding part of the southern hemisphere. Why?

Answers to Supplementary Problems

4.20. Sodium (Na^+) and chlorine (Cl^-) ions.

4.21. The average depth of the ocean basins is greater than the average height of the continents.

4.22. Organic debris such as the remains of plants and animals that live nearer the surface falls to the ocean bottom and serves as food for the creatures that live there.

4.23. About 10%; Antarctica.

4.24. Offshore; see Problem 4.14.

4.25. (a) The greater the wind velocity, the higher the waves. (b) The longer the period of time during which the wind blows, the higher the waves. (c) The greater the distance (*fetch*) over which the wind blows across the water, the higher the waves. Each of the above factors ceases to have a strong effect on wave height after a certain point; for example, after a day or two the waves will have reached very nearly the maximum height possible for the wind velocity and fetch of a given situation.

4.26. Below a depth of about half the wavelength of an ocean wave, there is almost no disturbance of the water, hence a submarine at such depths is in no danger.

4.27. Water particles move in approximately circular orbits as a wave passes by, so no net transport of water is involved in wave motion. Water waves do transport energy, as all other waves do.

4.28. (a) At first northward with the Gulf Stream, then northeastward with the North Atlantic Drift. (b) At first southward with the Canary Current, then westward and finally northwestward with the North Equatorial Current and the Gulf Stream.

4.29. The North Atlantic Drift, which is fed by the Gulf Stream, brings warm water to the shores of northwestern Europe. The prevailing westerlies that blow over this warm water lead to a moderate climate for this part of Europe. In the case of Labrador, the westerly winds have blown over the cold land mass of northern Canada; in addition, the waters that bathe the Labrador coast originate in Baffin Bay off Greenland and are accordingly very cold.

4.30. The middle latitudes of the northern hemisphere contain larger land masses than do those of the southern hemisphere, and land temperatures vary more than ocean temperatures.

Chapter 5

Minerals

CRYSTALS

Most solids are crystalline, with the atoms, ions, or molecules of which they consist being arranged in a regular pattern. Four kinds of bonds are found in crystals; ionic, covalent, metallic, and van der Waals.

In an *ionic bond,* one or more electrons from one atom are transferred to another atom, and the resulting positive and negative ions then attract each other. A crystal of ordinary salt, NaCl, is an example of an ionic solid, with Na^+ and Cl^- ions in alternate positions in a simple lattice (Fig. 5-1).

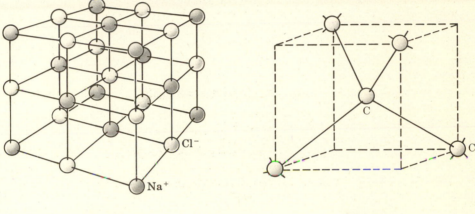

| Fig. 5-1 | Fig. 5-2 |

In a *covalent bond,* one or more pairs of electrons are shared by adjacent atoms. As these electrons move about, they spend more time between the atoms than elsewhere, which results in an attractive electrical force that holds the atoms together. The atoms in a molecule are held together by covalent bonds. An example of a covalent solid is diamond, each of whose carbon atoms is joined by covalent bonds to four other carbon atoms in a structure that is repeated throughout the crystal (Fig. 5-2). Both ionic and covalent solids are hard and have high melting points, which reflect the strength of the bonds. Ionic solids are much more common than covalent ones.

In a metal, the outermost electrons of each atom are shared by the entire assembly, so that a "gas" or "sea" of electrons moves relatively freely throughout it. The interaction between this electron sea and the positive metal ions leads to a cohesive force, much as in the case of the shared electrons in a covalent bond but on a larger scale. The presence of the free electrons accounts for such typical properties of metals as their opacity, surface luster, and high electrical and heat conductivities.

All molecules, and even inert-gas atoms such as those of helium, exhibit weak, short-range attractions for one another due to *van der Waals forces.* These forces are responsible for the condensation of gases into liquids and the freezing of liquids into solids even in the absence of ionic, covalent, or metallic bonds between the atoms or molecules involved. Such familiar aspects of the behavior of matter as friction, viscosity, and adhesion arise from van der Waals forces. Van der Waals forces arise from the lack of symmetry in the momentary distributions of the electrons in a molecule. When two molecules are close together, these momentary charge asymmetries tend to shift together, with the positive part of one molecule always near the negative part of the other even though the locations of these parts are always changing.   Van der Waals forces are quite weak, and substances composed of whole molecules, such as water, usually have low melting and boiling points and little mechanical strength in the solid state.

THE EARTH'S CRUST

The *crust* of the earth is its outer shell of rock.   The crust is typically 5 miles thick under the oceans and 20 miles thick under the continents.   The most abundant elements in the crust are oxygen (47% by mass) and silicon (28%); then come aluminum, iron, calcium, sodium, potassium, and magnesium, which range from 8% to 2% in that order.   Since the O^{--} ion is relatively large, over 90% of the volume of the crust is oxygen.

MINERALS

Rocks are aggregates of homogeneous substances called *minerals*.   Some rocks, for instance limestone, consist of a single mineral only, but the majority consist of several minerals in varying proportions.   Although over 2000 minerals have been identified, only a few are significant constituents of most rocks.   Among the commoner minerals or mineral groups are the feldspars, quartz, the ferromagnesian minerals, the clay minerals, and mica, which are all silicates, and calcite, which is composed of calcium carbonate.   The particular mineral or minerals that will be formed from a certain combination of elements depends upon the physical and chemical conditions under which the material crystallizes.

The silicate minerals, which make up about 87% of the earth's crust, consist of silicon combined either with oxygen alone or with oxygen and one or more metals.   The various silicates differ in composition and structure, and these differences result in a wide range of colors, hardnesses, crystal forms, and other properties.

SILICATE STRUCTURES

The basic structural element of all silicates is the SiO_4^{-4} tetrahedron in which each Si^{+4} ion is surrounded by four O^{--} ions that may be thought of as occupying the corners of a tetrahedron centered on the Si^{+4} ion (Fig. 5-3).   For clarity, the diagram is not drawn to scale; the O^{--} ions are actually quite close together, with the relatively small Si^{+4} ion just fitting in the central cavity.   The solid lines in Fig. 5-3 represent bonds between the Si^{+4} and O^{--} ions.   The bonds are partly covalent and partly ionic in character.   Each O atom thus has one electron free to form a bond with another atom.

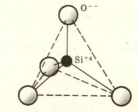

Fig. 5-3

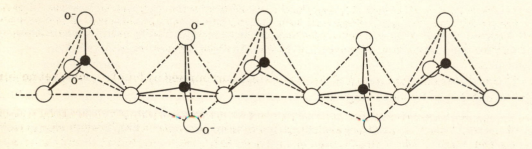

Fig. 5-4

In some silicates the SiO_4^{-4} tetrahedra occur singly, as in olivine where iron (Fe^{++}) and magnesium (Mg^{++}) ions hold them together.   In more complex silicates the tetrahedra are joined in single chains (as in augite) or double chains (as in hornblende) in which adjacent tetrahedra share O atoms (Fig. 5-4); the O^- ions at the top and sides of the chain can form ionic bonds with metal ions.   Sheets of tetrahedra are formed when each tetrahedron shares the three O atoms that form its "base" with neighboring tetrahedra (Fig. 5-5); the O^- ions at the top of the sheet can form ionic bonds with metal ions. Mica is an example of a mineral with a sheet structure.   Three-dimensional networks of SiO_4^{-4} tetrahedra are also possible in which all the O atoms of each tetrahedron are shared by its neighbors, as in quartz, SiO_2.

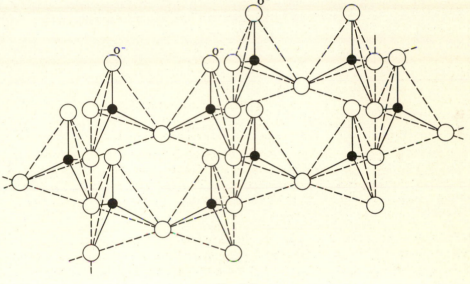

Fig. 5-5

Solved Problems

5.1.   Are all solids crystalline?

No.   The atoms, ions, or molecules of which a crystalline solid is composed fall into regular, repeated patterns. The presence of such long-range order is the defining property of a crystal.   Other solids lack long-range order in their structures, and may be regarded as supercooled liquids whose stiffness is due to an exceptionally high viscosity. Glass, pitch, and many plastics are examples of such amorphous ("without form") solids.

5.2.   Why can metals be deformed with relative ease whereas covalent and ionic solids are quite brittle?

Atoms in a metal can be readily rearranged in position because the bonding occurs by means of a sea of freely moving electrons.   In a covalent crystal the bonds are localized between adjacent atoms and must be ruptured to deform the crystal.   In an ionic crystal the bonding process requires a configuration of alternate positive and negative ions whose relative positions cannot be altered without breaking the crystal apart.

5.3.   In the silicate minerals each Si^{+4} ion is always surrounded by four O^{--} ions, yet no mineral has the formula SiO_4.   Why not?

An isolated SiO_4 tetrahedron has a net charge of -4 and so an assembly of such tetrahedra is electrically impossible.   The only mineral composed only of silicon and oxygen is quartz, SiO_2, in which adjacent tetrahedra share all of the O^{--} ions.   Minerals that contain isolated SiO_4^{-4} tetrahedra, such as olivine, also have positive metal ions in their crystal structures that bond these tetrahedra together, producing electrical neutrality.

5.4. The silicon-oxygen bond is a very strong one, hence the more oxygen atoms per SiO_4 tetrahedron that are shared by neighboring tetrahedra, the more stable the resulting mineral. On the basis of this information, which of the following minerals would you expect to be the most stable and which the least stable? Quartz, olivine, mica, feldspar.

In quartz all the oxygen atoms are shared between neighboring SiO_4 tetrahedra, so it is the most stable mineral of the ones listed. In olivine none of the oxygen atoms in the SiO_4 tetrahedra are shared, hence it is the least stable of these minerals.

5.5. The relative proportions of Fe^{++} and Mg^{++} in olivine are extremely variable. Why should this be so?

Fe^{++} and Mg^{++} ions have the same charge and nearly the same size (their radii are respectively 0.74×10^{-10} m and 0.66×10^{-10} m), and so are interchangeable in a given mineral structure. The relative proportions of iron and magnesium in a given olivine specimen depend upon the chemical and physical conditions under which the mineral was formed.

5.6. What property of Al^{+++} ions makes it possible for them to substitute for some of the Si^{+4} ions in many silicate minerals?

The two ions are nearly the same size (Al^{+++} has a radius of 0.51×10^{-10} m, Si^{+4} has a radius of 0.42×10^{-10} m) and so four O^{--} ions (radius 1.4×10^{-10} m) can fit as well around an Al^{+++} ion as around a Si^{+4} ion.

5.7. Certain silicate minerals, such as pyroxene, consist of chains of SiO_4 tetrahedra in which adjacent tetrahedra share an oxygen atom. (*a*) What is the ratio of silicon atoms to oxygen atoms in such minerals? (*b*) How is the requirement of electrical neutrality met by such minerals?

(*a*) Each SiO_4 tetrahedron shares two of its oxygen atoms with adjacent tetrahedra, so the chain contains three oxygen atoms per silicon atom.

(*b*) The SiO_3 chains are cross-linked by metal ions.

5.8. The feldspar minerals are abundant in the earth's crust. Describe the structures and compositions of the feldspars.

A feldspar consists of a three-dimensional assembly of SiO_4 tetrahedra in which some of the Si^{+4} are replaced by Al^{+++} ions plus other metal ions to maintain electrical neutrality. Thus in orthoclase each Al atom is accompanied by a potassium atom; the chemical formula of orthoclase is $KalSi_3O_8$. In the plagioclase feldspars sodium or calcium atoms or both are present to supplement aluminum atoms in replacing some of the silicon atoms. The compositions of these feldspars range from $NaAlSi_2O_8$ through various combinations of Na and Ca to $CaAl_2Si_2O_8$.

5.9. In certain minerals, such as beryl, rings of six Si atoms occur. (*a*) How many O atoms are associated with each ring? (*b*) What is the net charge of each ring? (*c*) In beryl, the silicate rings are linked by Be^{++} and Al^{+++} ions, in the ratio of three Be^{++} ions to two Al^{+++} ions. What is the chemical formula of beryl?

(*a*) There are 18 oxygen atoms in the ring configuration, with the six silicon atoms in the center of the six tetrahedra (Fig. 5-6).

(*b*) Since each ring consists of six Si^{+4} ions and 18 O^{--} ions, the total charge is -12.

(*c*) Each Be_3Al_2 group has a charge of $+12$, hence the chemical formula of beryl is $Be_3Al_2 \cdot Si_6O_{18}$.

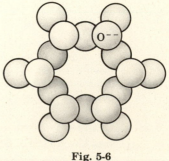

Fig. 5-6

5.10. The clay minerals are silicates of hydrogen and aluminum, some containing magnesium, iron, or potassium as well. They have layered structures in which each layer is electrically polarized, so that one side has a slight positive charge and the other a slight negative charge. Adjacent layers are held together by the attraction of the opposite charges that face each other; since the charges are small, the bonds between layers are feeble, and dry clay crumbles easily. Why does clay absorb water into its structure so readily?

Water molecules are polarized, with one end having a positive charge and the other a negative charge. Thus water molecules fit nicely between adjacent clay mineral layers, as in Fig. 5-7.

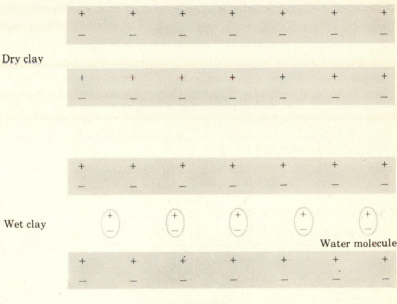

Fig. 5-7

5.11. Gold is found in native form and can therefore be considered as a mineral, but sodium is invariably found combined with other elements. What is the reason for this difference?

Gold is an inactive element, which means that it has little tendency to react chemically with other elements to form compounds. Sodium, on the other hand, is an extremely active element and combines readily with other elements, hence it is never found in the native state.

5.12. What is the *streak* of a mineral?

The streak of a mineral is the color of its powder. To determine its streak, a mineral sample is rubbed on a plate of unglazed porcelain. The streak of a mineral often varies less than the color of solid samples of that mineral and hence is useful in identification.

5.13. The hardness scale of minerals is as follows:

1. Talc (softest)
2. Gypsum
3. Calcite
4. Fluorite
5. Apatite
6. Orthoclase
7. Quartz
8. Topaz
9. Corrundum
10. Diamond (hardest)

How is this scale used to establish the hardness of a particular mineral sample?

Each mineral in the scale is harder than the ones of lower number and hence can scratch them, and in turn each one can be scratched by all those of higher number. Two minerals of the same hardness can scratch each other. The hardness of a sample therefore corresponds to that of the softest mineral on the scale that can scratch it. A fingernail is about 2.5 in hardness and a knife blade is about 5.5.

5.14. The *density* of a substance is its mass per unit volume; the density of a mineral is a characteristic property often valuable in identifying it. The SI unit of density is the kilogram per cubic meter (kg/m^3), so that the density of the aluminum, for example, would be expressed as 2700 kg/m^3. Another common unit of density is the gram per cubic centimeter (g/cm^3). Since 1 g/cm^3 = 1000 kg/m^3, the density of aluminum can also be given as 2.7 g/cm^3. The *specific gravity* of a substance is its density relative to that of pure water, which is 1000 kg/m^3 = 1 g/cm^3. The specific gravity of a substance is thus the same as the numerical value of its density when stated in g/cm^3, and is often used in place of density when identifying the substance. (*a*) A piece of quartz has a mass of 7.95 g and a volume of 3 cm^3. Find its density and specific gravity. (*b*) What would be the volume of a 1-g quartz sample? (*c*) What would be the mass of a 5-cm^3 quartz sample?

(*a*)   The density of quartz is

$$d = \frac{mass}{volume} = \frac{7.95 \text{ g}}{3 \text{ cm}^3} = 2.65 \text{ g/cm}^3$$

Hence its specific gravity is 2.65.

(*b*)   Since $d = m/V$, here

$$V = \frac{m}{d} = \frac{1 \text{ g}}{2.65 \text{ g/cm}^3} = 0.38 \text{ cm}^3$$

(*c*)   The basic formula $d = m/V$ now yields

$$m = dV = 2.65 \text{ g/cm}^3 \times 5 \text{ cm}^3 = 13.25 \text{ g}$$

5.15. What is meant by the cleavage of a mineral?

The cleavage of a mineral refers to any tendency it may have to split along smooth plane surfaces, which is determined by the nature of its crystal lattice. Cleavage occurs along planes in a crystal where the interatomic bonds are weaker than elsewhere. When a mineral grain is struck by a hammer, its cleavage planes (if any) are revealed as preferred directions of breaking. The existence of cleavage in a mineral is often obvious from the presence of flat faces in its grains. The number of cleavages in a mineral and the angles between them are useful in identifying the mineral. Mica, for instance, has perfect cleavage in one direction and readily splits into thin flakes, which reflects the weakness of the bonds between the silicate sheets of which it is composed compared with the bonds within each sheet. Halite (NaCl) has three perpendicular cleavages, as might be expected from its structure (Fig. 5-1). Quartz has no cleavage and, when struck, breaks along random curved surfaces like a piece of glass, which reflects the three-dimensional uniformity of its network structure.

5.16. Both cleavage and crystal form are characteristic mineral properties. What is the difference between them?

Crystal form refers to the shape of a crystal, which is determined by the pattern in which its constituent particles are linked together. Cleavage refers to the tendency, if any, of a crystal to break apart in a regular way, which is determined by the presence of weak bonds in certain directions in its structure.

5.17. Graphite consists of layers of carbon atoms in hexagonal arrays, with each atom covalently bonded to three others. The layers are bonded together by van der Waals forces. Would you expect graphite to exhibit cleavage?

Graphite exhibits cleavage because van der Waals bonds are much weaker than covalent bonds, hence the graphite layers can be readily broken apart. This is why graphite is so useful as a lubricant and in pencils.

5.18. Opal is an amorphous substance consisting of SiO_2 and H_2O in variable proportions. Would you expect opal to exhibit cleavage? How would you expect its specific gravity to compare with that of quartz, which is pure SiO_2?

Because opal is amorphous, it has no regular structure, and therefore cannot exhibit cleavage. The specific gravity of water is 1 and that of quartz is 2.65, hence a combination of SiO_2 and H_2O ought to have an intermediate specific gravity, and indeed that of opal ranges from 1.9 to 2.3.

5.19. What are the chief carbonate minerals?

The chief carbonate minerals are calcite, aragonite, and dolomite. Calcite and aragonite are both composed of calcium carbonate, $CaCO_3$, but have different crystal structures. Dolomite is similar to calcite but contains magnesium as well as calcium; its composition is usually represented as $CaMg(CO_3)_2$ although the proportion of Ca may vary.

5.20. Quartz and feldspar are both light in color and have a glassy luster. How could you distinguish a sample of quartz from one of feldspar?

Feldspar is softer than quartz and has two nearly perpendicular cleavage planes whereas quartz does not exhibit cleavage. Feldspar occurs in approximately rectangular crystals whereas quartz crystals are hexagonal and are transparent or translucent.

5.21. Gypsum and anhydrite both contain calcium sulfate, $CaSO_4$. What is the difference between them?

Water molecules are incorporated in the crystal structure of gypsum, so its composition can be represented as $CaSO_4 \cdot 2H_2O$. Anhydrite does not contain water and has a different crystal structure.

Supplementary Problems

5.22. Which of the fundamental interactions is involved in each of the various bonding mechanisms in molecules and solids?

5.23. What is the weakest type of bond found in solids? Give an example of a substance with this type of bonding.

5.24. What is the relationship between rocks and minerals?

5.25. What is the most abundant mineral in the earth's crust?

5.26. What gives asbestos its characteristic fibrous structure?

5.27. In quartz each SiO_4 tetrahedron shares all of its oxygen atoms with adjacent tetrahedra. What is the ratio between the numbers of oxygen and silicon atoms in quartz?

5.28. Olivine is a silicate mineral in which the SiO_4 tetrahedra do not share oxygen atoms with each other but are bonded together by iron and magnesium ions. (a) From this information would you expect olivine to exhibit cleavage? (b) Would you expect olivine to be light or dark in color?

5.29. A 50-g gold bracelet is dropped into a full glass of water and 2.6 cm³ of water overflows. What is the density of gold? What is its specific gravity?

5.30. Magnetite has a specific gravity of 5.2. (a) What is the mass of a piece of magnetite 1 cm × 2 cm × 10 cm? (b) What is the volume of a 1-kg piece of magnetite?

5.31. Magnetite, which consists of the iron oxide Fe_3O_4, and graphite, which consists of carbon, are both black minerals. Are there simple ways to distinguish them apart?

5.32. Which common minerals do not contain oxygen?

5.33. Quartz and calcite are similar in general appearance and are both commonly found in veins and on the sides of rock cavities. How can they be distinguished apart?

Answers to Supplementary Problems

5.22.   The electromagnetic interaction is responsible for all the bonding mechanisms in molecules and solids.

5.23.   Van der Waals; ice.

5.24.   A rock is an aggregate of grains of one or more minerals.

5.25.   Feldspar.

5.26.   Chains of SiO_4 tetrahedra linked by the sharing of O^{--} ions are responsible for the fibrous structure of asbestos.

5.27.   There are two O atoms per Si atom in quartz.

5.28.   (a) No cleavage; when struck, olivine breaks irregularly. (b) The presence of iron leads to a dark color.

5.29.   19 g/cm^3; 19.

5.30.   104 g; 192 cm^3.

5.31.   Magnetite is much harder than graphite and (as its name suggests) has magnetic properties.

5.32.   Oxygen is absent from halite (NaCl) and pyrite (FeS_2).

5.33.   Calcite crystals have perfect cleavage in three directions at angles of about 75°, whereas quartz crystals have no cleavage. Quartz is hard enough to scratch glass, but calcite is not. Unlike quartz, calcite dissolves in weak hydrochloric acid with effervescence.

Chapter 6

Rocks

IGNEOUS ROCKS

Rocks are classified as igneous, sedimentary, or metamorphic on the basis of their origins. *Igneous rocks* cooled from a molten state. Two-thirds of crustal rocks are igneous, with basalt constituting the bedrock under the oceans and granite the bedrock under the continents. *Intrusive* igneous rocks solidified beneath the surface where slow cooling resulted in large mineral grains; *extrusive* igneous rocks solidified after emerging from a volcano or other vent, and the rapid cooling resulted in small mineral grains. Some common igneous rocks are classified according to composition and character in Table 6-1.

Table 6-1

Composition	Intrusive Rocks	Extrusive Rocks
Quartz Feldspar (chiefly orthoclase) Ferromagnesian minerals, minor amount	Granite	Rhyolite
No quartz Feldspar (plagioclase) predominant Ferromagnesian minerals	Diorite	Andesite
No quartz Feldspar (plagioclase) Ferromagnesian minerals predominant	Gabbro	Basalt

SEDIMENTARY ROCKS

Most *sedimentary rocks* are composed of fragments of other rocks that have been eroded by the action of running water, glaciers, or wind. Shale, sandstone, and conglomerate are examples of such *clastic* (cemented fragment) rocks. Other sedimentary rocks consist of material that was either precipitated from solution or consolidated from the shells of marine organisms. Limestone, which consists largely of calcite, and chert, which consists largely of microcrystalline quartz, are examples. Sediments are usually deposited in layers, and the layering is evident in the resulting rocks. Although sedimentary rocks make up only about 8% of the crust, three-quarters of surface rocks are of this kind.

METAMORPHIC ROCKS

Metamorphic rocks were formed from igneous and sedimentary rocks under the influence of heat and pressure below the earth's surface. Sometimes the changes are in the minerals themselves, which turn into other varieties more stable in the different environment of the earth's interior; thus shale becomes slate or schist. Sometimes the changes are in the characters of the mineral grains, as in the metamorphism of limestone to marble or that of sandstone to quartzite.

Solved Problems

6.1.   How does the *Bowen reaction series* account for the different compositions of igneous rocks?

Molten rock underground is known as *magma*. As magma cools, the minerals that first crystallize may be converted into other minerals by reaction with the remaining liquid. N. L. Bowen discovered that a mafic magma (one rich in magnesium and iron) simultaneously undergoes two series of reactions during cooling. In the continuous reaction series, anorthite (calcium plagioclase feldspar) forms first. In the discontinuous reaction series, olivine is the initial mineral to crystallize. As the magma cools, much of the olivine dissolves and pyroxene minerals (single-chain ferromagnesian silicates) such as augite begin to form. If the mixture now cools quickly and hardens, the result will be the extrusive rock basalt or the intrusive rock gabbro, both composed largely of pyroxene and calcium feldspar with some olivine as well.

However, if the cooling is slow, the mineral crystals may settle out of the magma after they have come into being, leaving a progressively more silicon-rich liquid. The pyroxene that remains reacts with the cooling magma to yield amphibole minerals (double-chain ferromagnesian silicates) such as hornblende, while the plagioclase feldspar of the continuous series takes up more and more sodium into its structure. A hardening of the mixture at this point without further fractionation will yield the extrusive rock andesite or the intrusive rock diorite.

With continued slow cooling accompanied by the removal of newly formed mineral grains, biotite (black mica) appears in the discontinuous series while the continuous series progresses to albite (sodium plagioclase feldspar). Silicic rocks such as the extrusive rhyolite and the intrusive granite can be formed from a magma of this kind. Eventually both series converge in orthoclase (potassium) feldspar, and further fractionation yields muscovite (white mica) and ultimately, when only silica is left in the melt, quartz. The Bowen reaction series is shown in Fig. 6-1; the side arrows signify the removal of some of the intermediate minerals from the melt.

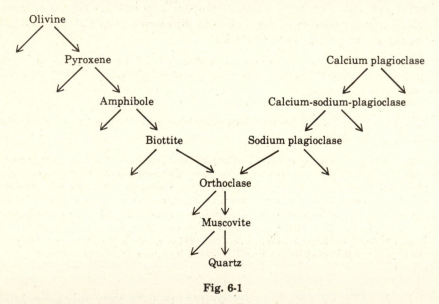

Fig. 6-1

6.2.   Discuss the validity of Bowen's concept of the origin of igneous rocks.

Some deposits of intrusive rock conform to the pattern suggested by Bowen, with olivine-rich rocks at the bottom and quartz-rich rocks at the top. In certain volcanoes, the sequence of lava flows is also consistent with Bowen's ideas, with basalt deposited first followed by andesite and finally small amounts of rhyolite.

There are a number of arguments against the general validity of Bowen's scheme. An important one is that no more than 5 or 10% of an original magma should end up as granite or granodiorite (a rock intermediate in composition between granite and diorite), yet enormous intrusive bodies of such rocks exist on the continents whereas intermediate rocks such as diorite, which ought to be much more abundant than granite, are scarce. Another objection is that the ocean floors and the various Pacific islands are composed of basalt with no sign of granite.

Probably actual magmas vary in their original composition, and processes other than differential crystallization (such as subsequent partial melting and recrystallization) also play a part in the formation of some granite bodies. Thus a magma that hardens into granite may be formed from the melting of sedimentary rocks well below the surface, rather than being the residue of an originally mafic magma, and some granite bodies may result from the intense metamorphism of sedimentary rocks without actual melting having occurred.

6.3. Igneous rocks rich in silicon are often called *silicic,* and rocks in which iron and magnesium are abundant are often called *mafic.* (*a*) Give examples of silicic and mafic rocks. (*b*) In general, silicic and mafic rocks are different in color. What is the difference? (*c*) Which are denser, silicic or mafic rocks? (*d*) Are silicic or mafic rocks more abundant in the earth's crust?

(*a*) SILICIC: granite and rhyolite; MAFIC: gabbro and basalt.

(*b*) Silicic rocks are light in color, mafic rocks are dark in color.

(*c*) Mafic rocks are denser.

(*d*) Mafic rocks are nearly twice as abundant in the crust.

6.4. Obsidian is a rock which resembles glass, in particular by sharing the property that its structure is closer to that of a liquid than to that of a crystalline solid. What does this observation suggest about the manner in which obsidian is formed?

To have the amorphous structure of a liquid, obsidian must have solidified so rapidly that crystals had no chance to develop. This can have occurred only by the cooling of a lava flow at the earth's surface.

6.5. Granite and rhyolite have similar compositions but granite is coarse-grained whereas rhyolite is fine-grained. What does the difference in grain size indicate about the environments in which each rock formed?

Large mineral grains can form only during slow cooling, hence granite must have solidified deep inside the crust. Small mineral grains occur when cooling is rapid, hence rhyolite must have solidified at or near the earth's surface.

6.6. *Lithification* refers to the hardening of sedimentary deposits into rock. What processes are involved in lithification?

Groundwater is able to circulate through coarse sediments and material precipitated from solution in the groundwater acts to cement the sediment grains. Common cementing materials are silica (which becomes quartz), calcium carbonate (which becomes calcite), and iron oxide. Cementation by silica leads to the hardest rocks. Some sediments have part of their original material dissolved away and replaced by other substances. Groundwater is unable to move freely through fine-grained sediments, which lithify primarily as the result of their compaction by the pressure of overlying deposits. During compaction, the particles of a sediment are squeezed together, which often causes some of their mineral crystals to grow larger and interlock. Slow chemical reactions in a compacted sediment may lead to the formation of new minerals, which may also contribute to lithification.

6.7. What are the most abundant sedimentary rocks? How does each originate?

In order of abundance, shale, sandstone, and limestone. Shale is formed from clay, sandstone from sand grains, and limestone from shell fragments or as a chemical precipitate.

6.8. What distinguishes the various clastic (cemented fragment) sedimentary rocks from one another?

The most common clastic sedimentary rocks are shale, sandstone, and conglomerate. They are distinguished according to grain size, from small (<1/16 mm) through medium (1/16 mm to 2 mm) to large (>2 mm) in the above order.

6.9. Give some examples of nonclastic sedimentary rocks. How are such rocks formed?

Limestone, chalk, diatomite, and chert are nonclastic sedimentary rocks. They may be formed by chemical or biological precipitation.

6.10.   Why is chert so resistant to chemical and mechanical attack?

Chert consists largely of microscopic quartz crystals and hence is hard and durable.

6.11.   What is *breccia*?

The term breccia is used to designate any rock that contains angular particles. Thus *sedimentary breccia* is a clastic sedimentary rock consisting largely of gravel-sized angular particles; conglomerate is similar but has rounded particles. *Volcanic breccia* consists of lava fragments that either stuck together while still hot or were cemented together in the same manner as a sedimentary rock. The rock fragments in a *fault breccia* originated during movement along a fault.

6.12.   The mineral grains of many metamorphic rocks are flat or elongated and occur in parallel layers, which gives the rocks a characteristic banded or layered appearance. (*a*) What is this property called?   (*b*) How does it originate?

(*a*)   *Foliation*.

(*b*)   Foliation occurs when the minerals of a rock recrystallize under great pressure, which, if the minerals have appropriate crystal structures, causes them to grow out perpendicular to the direction of the stress.

6.13.   Give examples of foliated and nonfoliated metamorphic rocks.

Foliated: slate, schist, gneiss.

Nonfoliated: marble, quartzite, hornfels.

6.14.   What happens to the density of a rock that undergoes metamorphism?

The density increases because the pressures under which metamorphism occurs lead to more compact rearrangements of the atoms in the various minerals.

6.15.   Gneiss is by far the most abundant metamorphic rock. Why? Describe the appearance of gneiss.

Gneiss is abundant because it can be formed from a wide variety of sedimentary, igneous, and other metamorphic rocks. It is coarse-grained, foliated, and often consists of layers of different mineral composition which lead to a banded appearance larger in scale than the foliation.

6.16.   Shale is a sedimentary rock that consolidated from mud deposits. What are the various metamorphic rocks that shale can become under progressively increasing temperature and pressure?

In order of increasing metamorphism, shale can become slate, schist, and gneiss.

6.17.   (*a*) What is the origin of limestone? (*b*) What rock does limestone metamorphose into?

(*a*) Limestone is produced both by consolidation of shell fragments and by precipitation of calcite (calcium carbonate) from solution.

(*b*) Marble.

6.18.   What is the difference between quartz and quartzite?

Quartz is a mineral whose chemical composition is SiO_2. Quartzite is a hard rock formed by the metamorphism of sandstone; it consists largely of quartz with micas, feldspars, and garnet also present.

6.19.   In what rock category does bituminous (soft) coal belong? Anthracite (hard) coal?

Bituminous coal can be considered as a sedimentary rock, anthracite as a nonfoliated metamorphic rock.

Supplementary Problems

6.20. Arrange the three classes of rock in order of their abundance in the crust.

6.21. In what type of rocks are fossils found?

6.22. Diorite is an intrusive rock and andesite, whose composition is similar, is an extrusive rock. How can they be distinguished apart?

6.23. What is generally true about the grain structures of rocks of volcanic origin?

6.24. What is the most common volcanic rock?

6.25. What are the two classes of sedimentary rocks?

6.26. What is chalk?

6.27. Limestone and marble have the same composition. What is the difference in structure that metamorphism produces when limestone becomes marble?

6.28. Gneiss is formed at much greater depths than slate. Which rock would you expect to have the greater density?

6.29. In what type of rocks is foliation found?

6.30. What are the two conditions necessary for a foliated rock to be formed?

6.31. What gives slate its characteristic shiny surfaces?

6.32. Which of the following rocks are hard, moderately hard, and soft? In each case indicate the nature of the rock.

Andesite	Gneiss	Obsidian
Chalk	Limestone	Quartzite
Chert	Marble	Shale

Answers to Supplementary Problems

6.20. Igneous, metamorphic, sedimentary.

6.21. Fossils are found in sedimentary rocks.

6.22. Diorite is coarse-grained and andesite is fine-grained.

6.23. Volcanic rocks cooled rapidly from a molten state and hence consist of fine grains.

6.24. Basalt is the most common volcanic rock.

6.25. (*a*) Fragmental (or detrital) rocks that consist of the fragments and decomposition products of other rocks. (*b*) Precipitates formed from material once dissolved in water and deposited either as a chemical precipitate or as the shells and bone fragments of dead organisms.

6.26. Chalk is a loosely consolidated type of limestone often consisting largely of shell fragments.

6.27. The grains become larger.

6.28. Gneiss.

6.29.   Metamorphic rocks exhibit foliation.

6.30.   (*a*) Such rocks are only formed from parent rocks under the influence of great pressure.   (*b*) At least some of the new minerals must have crystal structures that are elongated or flat; the growth of the mineral grains will be perpendicular to the applied stress.

6.31.   Minute flakes of mica.

6.32.   HARD: chert (sedimentary)
                                    quartzite (metamorphic)
            MODERATELY HARD: andesite (igneous)
                                    gneiss (metamorphic)
                                    limestone (sedimentary)
                                    marble (metamorphic)
            SOFT: chalk (sedimentary)
                                    shale (sedimentary)

Chapter 7

Weathering, Soil, and Groundwater

WEATHERING

The gradual disintegration of exposed rocks is called *weathering*. The chief mechanism of *mechanical weathering* is the freezing of water in crevices in rocks; water expands as it turns into ice, and considerable forces can be developed in this way. Surface water, which is slightly acid due to the formation of carbonic acid from dissolved CO_2 and to the presence of organic acids from decaying plant and animal matter, is the principal agent of *chemical weathering*. The weakly acid water turns feldspar into clay minerals, dissolves the calcite of limestone, and otherwise attacks many of the minerals of common rocks; quartz is the mineral that is least susceptible to chemical weathering.

SOIL

Soil is a mixture of rock debris, largely clay minerals and quartz fragments, and organic matter. Much of the latter consists of microorganisms such as bacteria and fungi and their remains. *Podzol soils* occur in moist, fairly cool climates under coniferous forests; *latosol soils* develop in tropical rain forests; *chernozem soils* are extremely fertile and were formed under prairie grasses in temperate, subhumid climates; and *desert soils* are found under arid conditions and contain little organic matter but many soluble minerals absent from soils in regions of abundant rainfall.

GROUNDWATER

The outer part of the earth's crust is permeable to water, and a considerable fraction of the rainwater reaching the ground is absorbed. The *water table* is the upper surface of the *saturated zone* in which the pore spaces of the rocks are filled with water. Groundwater moves slowly in the saturated zone and emerges in hillside *springs* where channels exist to the saturated zone and in streams, lakes, and swamps where the water table is above ground level. A *well* is a hole dug deep enough to reach to penetrate the water table and so make groundwater accessible.

The carbonate minerals of limestone and dolomite are soluble in groundwater made slightly acid by dissolved CO_2, and underground caverns are produced by the action of groundwater in regions containing such rocks. The roof of a cavern near the surface may collapse to form a *sinkhole;* a distinctive *Karst topography* occurs in places where underground channels, caverns, and sinkholes are abundant.

Groundwater contains traces of minerals in solution besides calcium carbonate, such as silica and ferric oxide, all of which, when precipitated, act as cementing materials in the hardening of sediments to form rock. Other processes also enter into *lithification,* such as the compaction of sediments by the pressure of overlying deposits and the recrystallization of certain of the minerals they contain. *Veins* are formed by the precipitation of various substances from groundwater in rock fissures.

Solved Problems

7.1. How are chemical and mechanical weathering related?

By breaking exposed rock into small fragments, mechanical weathering increases the surface area of a given volume of rock and so promotes the rate at which chemical weathering occurs. Chemical weathering can also lead to mechanical weathering, since many minerals increase in volume as they decay and thereby disintegrate the rocks in which they are incorporated.

7.2. In what way is the weathering of rock important to terrestrial life?

The rock debris produced by weathering is the principal constituent of soil.

7.3. Why are igneous and metamorphic rocks in general more susceptible to chemical weathering than sedimentary rocks?

Igneous and metamorphic rocks are formed under conditions of heat and pressure very different from those at the earth's surface, and minerals stable under the former conditions are not necessarily stable under the latter. In fact, the sequences of minerals in Fig. 6-1, which are arranged in descending order of temperature of formation, would be exactly the same if the arrangement were in the order of increasing stability against chemical weathering. Most sedimentary rocks, on the other hand, consist of rock debris that has already undergone chemical weathering, and so are relatively resistant to further attack. The chief exception is limestone, which is soluble in water that contains carbon dioxide.

7.4. Both marble and slate are metamorphic rocks. Would you expect a marble tombstone or a slate one to be more resistant to weathering?

Marble exposed to the atmosphere weathers fairly readily because its calcite content is soluble in rainwater that contains carbon dioxide. Slate consists largely of clay minerals that have metamorphosed to muscovite (white mica), which is nearly as resistant as quartz to chemical weathering.

7.5. What happens when ferromagnesian minerals such as olivine, the pyroxenes, and the amphiboles undergo chemical weathering?

These minerals all consist of SiO_4^{-4} tetrahedra bonded together by Mg^{++} (magnesium) ions and Fe^{++} (ferrous) ions. In the case of olivine, individual tetrahedra are bonded together in this way, in the cases of the pyroxenes and amphiboles, single and double chains of tetrahedra respectively are bonded together. During weathering, some Fe^{++} ions are oxidized to ferric oxide (Fe_2O_3) and some Mg^{++} ions go directly into solution. As a result, the crystal structures of these minerals collapse; the products of their decay are the clay minerals and minerals such as hematite that consist largely of ferric oxide.

7.6. When CO_2 is dissolved in water, some of it reacts with the water to form H^+ and HCO_3^- ions. The resulting slightly acid solution is commonly referred to as a "carbonic acid solution," although it is unlikely that any H_2CO_3 exists as such. Calcite ($CaCO_3$) reacts with "carbonic acid" to form the much more soluble calcium bicarbonate, $Ca(HCO_3)_2$. What is the formula for the overall reaction by which calcite goes into solution as calcium and bicarbonate ions?

$$CaCO_3 + H_2O + CO_2 \rightarrow Ca^{++} + 2HCO_3^-$$

7.7. Granite consists of feldspars, quartz, and ferromagnesian minerals. (a) What becomes of these minerals when granite undergoes weathering? (b) What kinds of sedimentary rocks can the weathering products form?

(a) The weathering of granite produces quartz fragments, clay minerals, and ferric oxides, plus K^+, Na^+, Mg^{++}, and Ca^{++} ions in solution.

(b) The quartz fragments can become sandstone or, when mixed with clay minerals, shale; the Ca^{++} and Mg^{++} ions can become limestone or dolomite.

7.8. What are the three principal layers, or *horizons,* of a typical soil?

The uppermost, or *A*, horizon consists of rock debris such as quartz sand, silt, and clay particles mixed with *humus,* which is partly decomposed plant debris.   It is rich in bacteria and fungi and their remains.   The *A* horizon is often referred to as *topsoil*.   The *B* horizon, or *subsoil,* is composed of mineral grains mixed with clay and also soluble materials such as iron oxide that have washed (or *leached*) out of the topsoil.   The *C* horizon is a region of partly disintegrated rock fragments that rests on the underlying bedrock.

7.9.   What is the origin of the red and yellow colors of some soils?

Red and yellow soil colors are due to staining by the iron oxides and hydroxides that result from the decomposition of ferromagnesian minerals.

7.10.   What is the distinction between the *porosity* and the *permeability* of a rock bed?

Porosity refers to the relative volume of pore space in a rock bed; the greater the porosity of a bed, the more water it can contain.   *Permeability* refers to the ease with which water can move through a rock bed; the greater the permeability of a bed, the faster water can flow through it.

7.11.   As a rule, clay and silt have greater porosity than gravel, but their permeability is less. Why?

A thin film of water tends to adhere to any solid surface through forces of molecular attraction (these are an example of the van der Waals forces mentioned in Chapter 5).   When two surfaces are very close together, the adhesive forces can extend through the entire space between them and hold any water there in place.   In clay and silt, the grain size and hence the pore spaces are very small, and as a result their permeability is low.   In gravel the grains are sufficiently large for the pore spaces to be wider than the thickness of the water films that adhere to the grains, so that water can move through a gravel bed relatively freely.

7.12.   Draw a cross-sectional view of a region underlain by permeable rock and show the water table, a well, a lake, and a spring.

See Fig. 7-1.

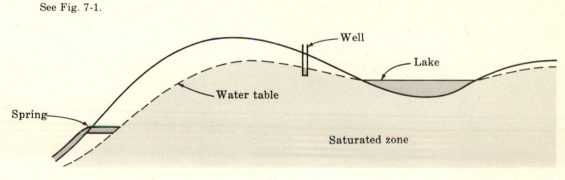

Fig. 7-1

7.13.   An *aquifer* is a body of permeable rock or sediment through which groundwater moves.   In an artesian well or spring, water rises above the aquifer that feeds it.   What geological conditions are required for such a situation to occur?

An artesian system can exist when an inclined aquifer is sandwiched between impermeable layers, as in Fig. 7-2.   If the level of the water table in the aquifer is sufficiently high, water will be forced to the surface either through a fissure in the upper impermeable layer to form an artesian spring or through a well dug down to the aquifer.

7.14.   What is "hard" water?   How can it be made "soft"?

"Hard" water contains dissolved minerals which prevent soap from forming suds, react with soap to produce a precipitate, and form insoluble deposits in boilers.   Calcium and magnesium ions are usually responsible for hard water. Groundwater often contains these ions through the solvent action of water containing dissolved CO_2 on rocks such as limestone.   To soften water, the Ca^{++} and Mg^{++} ions must be removed, which can be done in a variety of ways.   In one common method, hard water is passed through a column containing a mineral called zeolite, which absorbs Ca^{++} and Mg^{++} ions into its structure while releasing an equivalent number of Na^+ ions. Since Na^+ ions do not affect soap, nor do sodium compounds precipitate out from hot water, the water is now soft.

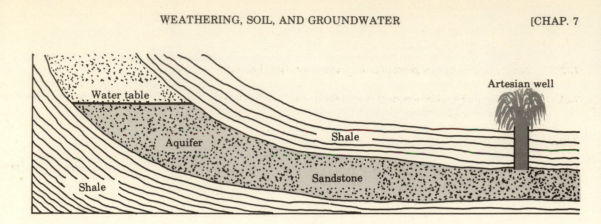

Fig. 7-2

7.15. What is *evapotranspiration* and why is it important in the water budget of a region?

   Evapotranspiration refers to the return to the atmosphere of water from the land by direct evaporation and by transportation by plants, which is the process by which water is drawn into a plant through its roots and given to the atmosphere through its leaves. If annual evapotranspiration in a region exceeds the water deposited annually by precipitation, the water needed for normal plant growth will have to be provided by irrigation.

Supplementary Problems

7.16. What common rocks are almost immune to chemical weathering?

7.17. Why are clay minerals and quartz particles abundant in sediments which have not been chemically deposited?

7.18. Where would you expect to find Karst topography?

7.19. What is the chief factor that determines what kind of soil will develop in a given region?

7.20. Would you expect sedimentary or igneous rocks to have the greater porosity?

7.21. Where does the water come from that feeds a spring?

7.22. Which of the following is typical of the speeds with which groundwater moves through an aquifer? One ft/hour; 1 ft/week; 1 ft/year.

7.23. Why are mineral deposits near hot springs thicker than those near ordinary springs?

Answers to Supplementary Problems

7.16. Quartz is highly resistant to chemical attack, hence rocks which are largely quartz, such as chert and many quartzites, are not subject to chemical weathering.

7.17. Quartz is resistant to chemical attack and so survives weathering. Feldspar, the most common mineral, is converted into clay minerals by the "carbonic acid" of surface waters.

7.18. Karst topography is found in regions of abundant rainfall that are underlain by carbonate rocks such as limestone and dolomite.

7.19. The climate of the region.

7.20 Sedimentary rocks in general have greater porosity than igneous rocks.

7.21. A spring is fed by groundwater that reaches the surface in a more or less definite channel. The groundwater itself was orginally rainwater.

7.22. One ft/week.

7.23. Minerals are more soluble in hot water than in cold water, hence the mineral content of water from a hot spring is greater than that of water from a cold spring.

Chapter 8

Erosion

STREAMS

The most powerful agent of erosion is the running water of streams and rivers, whose work is aided by the pebbles and stones carried in their flow. When the slope of a stream is steep, it carves a narrow V-shaped valley, with the debris being transported away by the swiftly moving water. At lower elevations where the slope is more gradual, the stream tends to widen rather than deepen its bed, and some of its load of *alluvium* (erosional debris) is deposited on the broad *flood plain* of the resulting valley. A *delta* may be formed from alluvium where a stream empties into a lake, sea, or ocean. Deposition is also common where a stream leaves a steep valley and slows down as it enters a plain; the resulting cone of sand and gravel pointing upstream is called an *alluvial fan*. The most widespread accumulations of sediments are found in the shallow parts of the oceans adjacent to the continents.

SEDIMENTS

The physical characteristics of sediments often reveal something about their origin. A smoothly flowing stream tends to deposit first the coarse fragments it carries, with finer and finer particles settling out farther and farther along its path. Thus a *well-sorted* sediment, which consists of particles of very nearly the same size, indicates that such a stream was responsible. On the other hand, a turbulent stream usually leaves a *poorly-sorted* sediment composed of particles of all sizes mixed together.

The type of *bedding,* or layering, of a sediment also provides clues as to the conditions under which it was formed. Slight changes in the composition or grain size of a sediment produce layers of different appearance, which may lie parallel to one another or at different angles. The latter situation, called *cross-bedding,* can arise from turbulent flow which produces patterns of ripples on the channel bottom; when the angles between the beds are steep, the deposits are more likely to have been laid down by winds in the form of dunes than by streams.

GLACIERS

A *glacier* is a large mass of ice that has formed from the recrystallization under pressure of accumulated snow. *Valley glaciers* occur in mountain valleys originally cut by streams, and flow slowly downhill to melt at their lower ends. A valley glacier grinds out a characteristic U-shaped trough with a round, steep-walled *cirque* at its head. The piled-up debris at the foot of a glacier is called a *moraine.*

Continental glaciers, or *icecaps,* that are thousands of feet thick cover most of Greenland and Antarctica; motion in such an icecap is from the center outward, and icebergs are fragments that have broken off the edges into the sea. Similar sheets of ice extended across Canada and northern Eurasia in relatively recent geological history.

WIND AND WAVES

In desert regions wind acts as an erosional agent by virtue of the sand it carries along the surface, but its effects are minor compared with those of the flash floods caused by the occasional rain storms. Wind is significant in shaping certain landscapes for another reason, its ability to transport fine silt for considerable distances while leaving behind dunes of heavier sand particles. Deposits of windblown silt are called *loess*.

Ocean waves battering a coastline gradually wear it away to form beaches and cliffs, but, as in the case of winds, the more important effect of their activity lies in the redistribution of sediments produced by other agents.

Solved Problems

8.1.   What is the source of energy that makes possible the erosion of landscapes?

The ultimate source of most of the energy that goes into erosion is the sun. Solar energy evaporates surface water, some of which subsequently falls as rain and snow on high ground. The gravitational potential energy of the latter water turns into kinetic energy as it flows downhill, and some of the kinetic energy becomes work done in eroding landscapes in its path. Winds and waves, too, derive their energy from solar radiation.

8.2.   What is the eventual site of deposition of most sediments?

The ocean floor.

8.3.   The downward motion of rock debris and soil under the influence of gravity alone is known as *mass wasting*. What are the chief varieties of mass wasting?

Soil and other weakly-consolidated deposits move downslope by the gradual process of *creep*. Much more rapid but intermittent in occurrence are the various kinds of landslide: *rockfalls* and *rockslides,* in which rock masses break off a cliff or steep slope; *slumps,* in which large bodies of rock or soil slip downward along a concave slope; and *debris flows,* in which large volumes of rock debris slide downward suddenly rather than gradually as in the case of creep.

8.4.   Distinguish between the laminar and turbulent flow of a stream. Which is more effective in transporting sediments?

In laminar (or "streamline") flow, each particle of water passing a particular point follows the same path as the particles that passed that point before. The direction in which the individual particles move is the same as that in which the stream as a whole is moving.

In turbulent flow, the water motion is irregular. Successive water particles passing the same point do not in general follow the same paths, and in places they move opposite to the direction in which the stream as a whole is moving. Eddies and whirlpools are characteristic of turbulent flow. Turbulence occurs when the stream velocity is high, when the stream bed is rough, and when there are obstructions or sharp bends; mountain streams commonly exhibit turbulent flow. Sediments have a greater tendency to remain suspended in a turbulent stream, hence such a stream is more effective in transporting them.

8.5.   What are the various ways in which a stream can transport the debris of erosion?

Some material is carried in solution. Small particles can be carried in suspension in a turbulent flow; the greater the turbulence, the larger the particles that can be carried. Even a relatively smoothly-flowing stream will be turbulent near its bed owing to irregularities there. Particles too large to be part of the suspended load of a stream can be pushed along its bed by the force of the moving water, either in a succession of jumps ("saltation") or by rolling or sliding.

8.6.   Some stream patterns consist of irregular branches in all directions, much like the branches of an oak tree (hence the term *dendritic*), whereas others consist of branches that are more or less parallel, much like vines on a garden trellis. What determines which of these stream patterns will develop in a particular region?

A dendritic stream pattern develops where the surface rocks are uniformly resistant to erosion. A trellis pattern develops where the edges of tilted rock strata whose resistance to erosion differs are exposed, with the weaker rocks being preferentially cut to form stream beds (Fig. 8-1).

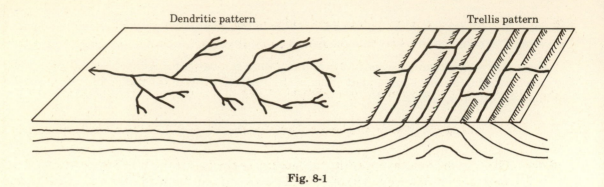

Fig. 8-1

8.7. Trace the evolution of the landscape of an initially uplifted region in which stream erosion is the chief geological process. Why is this cycle no longer believed to be widely applicable in its simplest form?

In the young landscape, streams have steep gradients and cut narrow, deep valleys in the predominantly high land mass. As time goes on, tributary streams develop and the region is carved into an intricate pattern of ridges and valleys. This mature landscape is eventually worn down into a series of broad flood plains covered with alluvium and separated by low hills. With old age the landscape becomes a low, rolling *peneplain* near sea level.

The main reason the above cycle, though useful in classifying landscapes, has fallen out of favor with geologists is that there are very few regions in which geological processes involving uplift do not occur at the same time as stream erosion. Thus most actual landscapes are the result of a complex of different factors, and reflect a balance among them rather than the action of stream erosion alone.

8.8. Why do meanders form in a river flowing along a flood plain?

A meander begins when the river current is directed against one bank by some obstruction or irregularity in its bed. The bank is then eroded in that place to form a concavity. The concavity directs the current against the opposite bank farther downstream to form another concavity there (Fig. 8-2). The eroded material is deposited where the water speed is least, which is opposite the concavities in the banks; these places become convex, and the result is a series of bends as the river swings from side to side. Thus the formation of meanders is inevitable where a river flows through a flood plain.

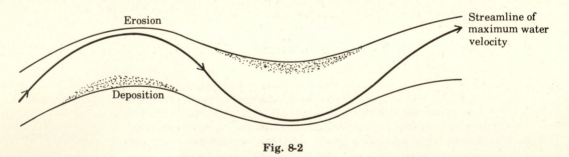

Fig. 8-2

8.9. A *hanging valley* is a tributary valley that is higher than its main valley where they meet, with the tributary stream joining the main stream via a waterfall. How does a hanging valley originate?

Hanging valleys occur in glacially eroded valleys. As a rule the main valley is cut more deeply than its tributary valleys, hence when the ice melts the tributaries are higher.

8.10. What properties would you expect to find in the sediments deposited as an alluvial fan where a steep stream emerges onto level ground?

The deposited material will consist of irregularly bedded sand and gravel or clay, along with rounded pebbles and stones.

8.11. Under what circumstances does a glacier form?

A glacier forms when the average annual snowfall in a region exceeds the annual loss by evaporation and melting.

8.12. Glaciers are observed to wear down bedrock that is harder than glacial ice. How can this happen?

Embedded in glaciers are stones and boulders, some of which are hard enough to erode the bedrock.

8.13. Glaciers grind away rock with far more force than rivers or streams, yet running water has had more influence in shaping landscapes around the world than glaciers have. Why?

Glaciers were insignificant or absent during most of the earth's history except for relatively brief "ice ages"; even today glaciers are active over only a small (10% or so) proportion of the earth's land area.

8.14. How can a valley formed by glacial erosion be distinguished from one formed by stream erosion?

A glaciated valley has a U-shaped cross section, as compared with the V-shaped cross section of a stream valley. At the head of a glaciated valley is a round bowl-shaped hollow with steep walls called a *cirque*, as compared with the progressively narrowing river bed of a stream valley. Rock fragments carried along by the glacier in its motion polish and gouge bedrock in its path to leave characteristic smooth, grooved surfaces.

8.15. Which agent of erosion was chiefly responsible for the ruggedness of the mountain landscapes characteristic of the Rockies, the Alps, and the Himalayas? How does the activity of this agent today compare with what it was in the past?

Valley glaciers produced the deep gorges, steep slopes, and sharp ridges of these mountain ranges. The glaciers were larger and more numerous in the past when climates were more severe than today.

8.16. Unstratified material deposited by a glacier is called *till*. What are the distinctive properties of till?

Till is a mixture of debris ranging from fine, claylike material to large boulders. Till is neither sorted in size nor deposited in distinct layers as other sediments are. Some of the boulders exhibit flattened, scratched faces where they were scraped against bedrock.

8.17. Why do flash floods occur in deserts and why are they so important in forming desert landscapes despite their infrequency?

In a desert region there is little soil or vegetation to absorb water from a rainfall, so most of it remains on the surface to flood channels that are dry at other times. The rapid flood waters are effective erosional agents and carry considerable loads of sediments, which are often deposited in alluvial fans at the feet of desert mountain ranges since no permanent watercourses exist to carry them to the sea. The flood waters themselves eventually evaporate. Wind is much less able to affect landscapes than running water, hence desert landscapes reflect the action of flash floods to a greater extent than that of winds.

8.18. The minerals in loess—mainly quartz, feldspar, calcite, and mica—show little chemical weathering, unlike minerals in most other sedimentary deposits. Why?

The silt particles deposited by winds to form loess were originally present in glacial drift or in sediments produced by mechanical weathering in desert regions. In both cases the particles had little exposure to liquid water, and hence little chemical weathering occurred.

8.19. What characteristics of a sedimentary rock would suggest an arid climate at the time the original sediments were deposited?

The presence of well-sorted, rounded sand grains, the absence of clay and gravel, and cross-bedding in large, sweeping curves are characteristic of sediments deposited by winds in desert regions.

8.20.   What is a *turbidity current*?

A turbidity current involves the motion of a dense mixture of water and suspended sediments along the floor of a lake, sea, or ocean.   Turbidity currents are thought to be one of the processes responsible for the sorting of ocean sediments to form graded layers.

Supplementary Problems

8.21.   Under what climatic conditions would you expect erosion to be most rapid?   Least rapid?

8.22.   What is *talus*?

8.23.   What agent of erosion produces valleys with a V-shaped cross section?   A U-shaped cross section?

8.24.   Distinguish between a *delta* and an *alluvial fan*.

8.25.   Which of the following is typical of the speed with which a valley glacier moves downhill?   One ft/hour; 1 ft/day; 1 ft/month.

8.26.   Moraines are common in the North Central states.   What does this suggest about the recent geological history of this region?

8.27.   What are the two general types of wind-deposited sediments?

8.28.   What mineral or minerals are most abundant in sand derived from the wave erosion of granite?

8.29.   What is the probable origin of a thick, evenly-bedded limestone?

Answers to Supplementary Problems

8.21.   Erosion is most rapid in a hot, moist climate and is least rapid in a cold, dry climate.

8.22.   Talus refers to rock debris at the foot of a cliff from which rockslides or rockfalls have occurred.

8.23.   Streams; glaciers.

8.24.   A delta is formed from alluvium deposited where a stream flows into still water, for instance a lake or sea.   An alluvial fan is formed from alluvium deposited where a stream slows down as its slope decreases.

8.25.   One ft/day.

8.26.   Ice sheets advanced south to this region from centers of accumulation in Canada at various times in the past; the most recent glaciation began to recede only about 20,000 years ago.

8.27.   (*a*) Sand; (*b*) loess, which is composed of fine silt.

8.28.   Quartz.

8.29.   Such a deposit is probably the result of the precipitation of calcite from groundwater.

Chapter 9

Vulcanism and Diastrophism

VOLCANOES

Molten rock is called *magma* when it is underground and *lava* when it is on the surface. A *volcano* is an opening in the earth's crust through which lava, rock fragments, and hot gases emerge. Since lava cools and hardens relatively near the opening, a characteristically conical mountain is usually built up. Most volcanoes are only intermittently active. Explosive eruptions occur when the magma is highly viscous and has a large gas content; a less viscous magma with a small gas content flows out more or less quietly.

Volcanic rocks are fine-grained because lava cools too rapidly for large mineral crystals to form, and often they contain holes where gas bubbles were trapped; *pumice* is an extreme example of the latter effect. In explosive eruptions bits of magma are blown out which solidify into fragments of various sizes collectively called *pyroclastic debris*. Deposits of the finer fragments may consolidate into *tuff,* and deposits of the coarser ones into *volcanic breccia.*

The majority of today's volcanoes occur in a band that encircles the Pacific Ocean and in another that extends from the Mediterranean region across Asia to join the Pacific band in the Indonesian archipelago.

PLUTONS

A *pluton* is a body of magma that has risen through the crust and hardened while still underground. Plutons consist of coarse-grained igneous rocks. A *batholith* is a large pluton that may be several miles thick and extend over thousands of square miles; most of the rock in batholiths is granite. Batholiths are always associated with mountain ranges, either past or present. *Sills* and *laccoliths* are smaller plutons that have intruded parallel to existing strata; *dikes* are wall-like plutons that have intruded into fissures that cut across existing strata (Fig. 9-1).

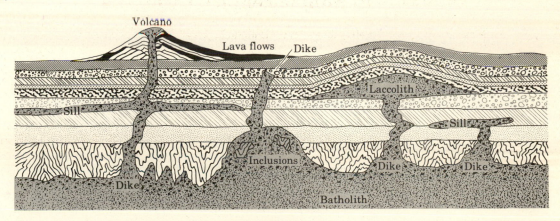

Fig. 9-1

DIASTROPHISM

Diastrophism refers to movements of the solid rock of the earth's crust. An example is the relative displacement of the rocks on both sides of a fracture; a fracture along which such a slippage has occurred is called a *fault*.   The commonest cause of earthquakes is the sudden dislocation of rocks along a fault.   Sometimes a crustal segment subjected to a stress may flow instead of faulting; thus folds in rock strata result from gradual yielding to a horizontal compression.

MOUNTAIN BUILDING

Most of the major continental mountain ranges, such as the Rockies, the Alps, and the Himalayas, are the result of large-scale folding.   Smaller mountain ranges have often been produced by faulting which elevated crustal blocks; volcanic action is also responsible for some mountain ranges.

The sedimentary layers in fold mountains typically increase in thickness toward the center of the range, which suggests that where the range now stands was once a large basin that gradually sank as sediments accumulated in it.   Such a sinking basin is called a *geosyncline*.   Eventually diastrophic activity, probably connected with continental drift (Chapter 11), led to the folding and uplifting of the sedimentary deposits, which were later intruded by granite batholiths and other plutons.   It is thought that the granite came into being when the base of the geosyncline descended far enough to melt and the granitic magma then rose to intrude the overlying strata.

Solved Problems

9.1.   Do all volcanic eruptions lead to the production of conical mountains?

Quiet eruptions may result in the formation of a broad dome with gradual slopes called a *shield volcano*. The volcanoes that make up the island of Hawaii are of this kind.   Large-scale lava flows have occurred in the past which hardened into huge sheets of *flood basalt* or *plateau basalt,* some of them hundreds of thousands of square km in area and hundreds or thousands of meters thick.   The Columbia Plateau in the northwestern United States came into being in this way, as did a region of the North Atlantic that includes southern Greenland, Iceland, northern Ireland, and the islands of western Scotland.   Only low-viscosity basaltic lava is able to cover such large areas.

9.2.   A *caldera* is a large craterlike depression up to several miles across that is associated with former volcanic activity.   How is a caldera formed?

A caldera can be formed in two ways, or by a combination of both: a violent explosion that ejects a vast quantity of material to leave behind a depression; and the collapse of a volcanic cone into an underground cavity left by the loss of magma, either through being blown out or by seeping away.

9.3.   What are the chief constituents of volcanic gases?

Steam comprises most of the gases emitted by a volcano.   Some of the steam comes from groundwater heated by magma, some comes from the combination of hydrogen in the magma with atmospheric oxygen, and some (called *juvenile water*) was formerly incorporated in rocks deep in the crust and is carried upward by the magma to be released at the surface.   A typical magma probably contains about 2% juvenile water.   Other gases include carbon dioxide, nitrogen, sulfur dioxide and trioxide, chlorine, carbon monoxide, hydrogen, argon, methane, and hydrochloric and hydrofluoric acids in variable proportions.

9.4.   What kind of rock is usually the principal constituent of the lava flow from an explosive volcano?   From a quiet volcano?

Explosive eruptions usually involve molten rhyolite, which is very viscous and hence prevents gas from escaping readily.   Quiet eruptions usually involve molten basalt, which is quite fluid and permits gas to escape before pressure is built up.

9.5.   If you wished to investigate the properties of a particular kind of magma, would it be sufficient to melt a sample of rock that had hardened from such a magma and study it in the laboratory?

No. Every magma contains volatile material such as water whose presence significantly affects its properties.

9.6.   What is the chief factor that determines the viscosity of a magma, that is, how readily it flows? What kinds of landscapes are produced by volcanoes whose lavas have relatively high and relatively low viscosities?

The greater the silicon content of a magma, the higher its viscosity and the less readily it flows. Highly viscous lavas usually produce steep conical mountains and, in general, a rugged landscape; less viscous lavas spread out to produce more even landscapes.

9.7.   What is the cause of the holes found in many volcanic rocks such as pumice?

Such holes were produced by bubbles of gas trapped in lava as it solidified.

9.8.   What is the origin of the glasslike rock obsidian?

Obsidian forms from a rhyolitic lava, which is rich in silica, that cools so rapidly that crystals do not have time to grow. The resulting rock, like glass, has a structure that is essentially that of a liquid.

9.9.   A *geyser* is a hot spring from which hot water and steam are violently ejected from time to time. Describe the physical process that is believed to be responsible for geyser eruptions.

The variation of the boiling point of water with pressure is believed to underlie the behavior of a geyser. The higher the pressure, the higher the boiling point; this effect is employed in a pressure cooker to enable food to be heated in water to temperatures above 100 °C. A geyser consists of a series of irregular subsurface passages filled with groundwater that lead up to a vent at the surface. The pressure in the water increases with depth, and the boiling point increases accordingly; it is about 120 °C at a depth of 10 m and about 200 °C at a depth of 150 m. The water in the geyser passages is heated by hot gases or by contact with hot rock until it is all nearly boiling. When boiling begins to occur, steam forces some water out of the vent, which reduces the pressure on the water remaining in the underground passages. The sudden release of pressure then causes much of the superheated water to turn into steam all at once, and a column of steam and boiling water shoots out. The empty underground passages then gradually refill with water and the cycle repeats itself.

9.10.   Distinguish between a *dike* and a *vein*.

A dike consists of molten rock that has intruded into a fissure and hardened there; a vein consists of material that has precipitated in a fissure from solution in groundwater.

9.11.   What kinds of rocks are likely to be found in (*a*) a batholith and (*b*) a dike?

(*a*)   Because magma cools slowly in a batholith, coarse-grained igneous rocks such as granite, diorite, and gabbro are likely to be found.

(*b*)   Magma may cool slowly or rapidly in a dike, depending on the circumstances. Hence both coarse- and fine-grained igneous rocks may be found: granite, diorite, gabbro, rhyolite, andesite, and basalt, for instance.

9.12.   What effect does the intrusion of a batholith have on the nature of nearby rocks?

The initial high temperature of a batholith, which persists for a long time since it cools slowly owing to its great mass, produces changes in nearby rocks that constitute *contact* (or *thermal*) *metamorphism*. In general, mineral grains grow larger in such rocks and become interlocked. These changes are especially marked in soft sedimentary rocks; thus sandstone becomes quartzite, shale becomes hornfels, and limestone becomes marble. Contact metamorphism produces unfoliated rocks since only heat is involved and not pressure as well.

9.13.   Would you expect to find a wider region of contact metamorphism near a dike or near a batholith?

Near a batholith, because of the greater amount of heat that had to be dissipated in its cooling.

9.14. Show the relationships among sedimentary, igneous, and metamorphic rocks, unconsolidated sediments, and magma by means of a diagram.

See Fig. 9-2.

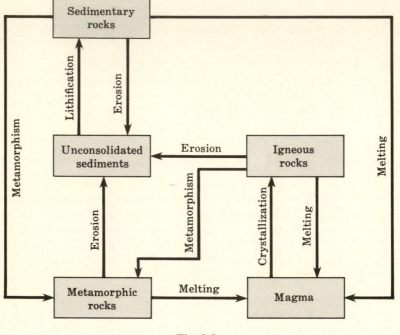

Fig. 9-2

9.15. A fracture in a rock may be either a *joint* or *fault*. What is the difference between them?

A joint is a fracture surface along which no movement has taken place, whereas relative motion between the rocks on either side of the fracture surface has occurred in the case of a fault. A joint may be caused by the contraction of molten rock as it cools, as well as by mechanical stresses in the crust.

9.16. What are the principal classes of faults?

A *normal fault* is an inclined surface along which a rock mass has slipped downward (Fig. 9-3a). A normal fault is the result of tension in the crust and increases the area covered by the rocks involved.

A *thrust* (or *reverse*) *fault* is an inclined surface along which a rock mass has moved upward to override the neighboring mass (Fig. 9-3b). A thrust fault is the result of compression in the crust and decreases the area covered by the rocks involved.

A *strike-slip fault* is a surface along which one rock mass has moved horizontally with respect to the other (Fig. 9-3c). A strike-slip fault is the result of oppositely-directed forces in the crust that do not act along the same line, so that the result is a distortion of the crust rather than a change in area.

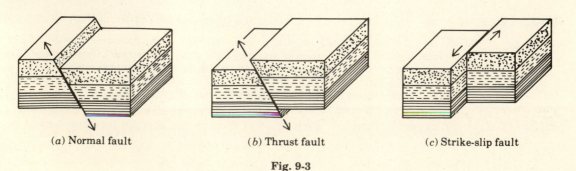

(a) Normal fault   (b) Thrust fault   (c) Strike-slip fault

Fig. 9-3

9.17. **What topographical features are associated with faults?**

Both normal and thrust faults produce cliffs called *fault scarps*. A strike-slip fault is often marked by a *rift*, which is a trench or valley caused by erosion of the disintegrated rock produced during the faulting.

9.18. **Distinguish between a *syncline* and an *anticline* in folded rock strata.**

A *syncline* is a troughlike fold that is concave upward; an *anticline* is an archlike fold that is convex upward (Fig. 9-4).

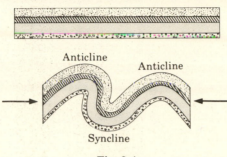

Fig. 9-4

9.19. **What kind of faults would you expect to find in a region that has undergone severe folding?**

Folding is the result of horizontal compressional forces in the earth's crust which have deformed existing rock strata. If the rocks are not sufficiently plastic they will fracture rather than fold, and the result will be thrust faults where one crustal block rides up on another to reduce the area of that region.

9.20. **What are the four chief kinds of mountains?**

(a) *Fold mountains* are the most numerous and consist of sedimentary deposits that accumulated in geosynclines and were folded, faulted, and uplifted. Plutons and metamorphic rocks are common in the cores of fold mountains. The Rockies, the Alps, and the Himalayas are folded mountain ranges.

(b) *Fault-block mountains* were formed by the elevation of large sections of the crust and are bounded by normal faults. Unlike fold mountains, their histories vary considerably: those that consist of geosynclinal rocks may or may not have undergone folding earlier, and may or may not have been deeply eroded prior to the faulting; other fault-block mountains consist of lava flows. The Sierra Nevada of eastern California is a fault-block range that consists of a crustal segment roughly 600 km long and 100 km wide that was raised and tilted westward.

(c) *Upwarped mountains* are the result of an elevation of a region of the crust not associated with lateral compression, as in the case of fold mountains, or with tension, as in the case of fault-block mountains. The Adirondack Mountains of New York State consist of an ancient mountain range that had been completely leveled by erosion long ago and only recently uplifted.

(d) *Volcanic mountains* are most common in the oceans, where they constitute island arcs such as the West Indies and Aleutians and midocean ridges such as the Mid-Atlantic Ridge. Among the relatively few continental volcanic mountains are the Cascade Range in Washington and Oregon, Kilimanjaro and Kenya in Africa, and a number of Andean peaks in South America.

9.21. **Why is it believed that the region where the Rocky Mountains now stand was once near or below sea level?**

The Rocky Mountains contain thick layers of sedimentary rocks that can only have been formed from sediments deposited over a long period of time. Hence the region must once have been low enough for rivers and streams containing erosional debris to flow into it.

9.22. **What is the difference between right-lateral and left-lateral strike-slip faults?**

When facing a strike-slip fault, if the farther side of the fault has moved to the right, it is a right-lateral fault. If the farther side has moved to the left, it is a left-lateral fault. The strike-slip fault shown in Fig. 9-3(c) is a left-lateral one.

Supplementary Problems

9.23.   What kinds of rocks are likely to be found in lava flows?

9.24.   What is the most common volcanic rock?

9.25.   What causes the hardening of deposits of pyroclastic debris into tuff and volcanic breccia?

9.26.   Distinguish between concordant and discordant bodies of intrusive rock.   Give examples of each class.

9.27.   Why do dikes and sills usually consist of fine-grained rocks?

9.28.   Why are metamorphic rocks often found near plutons?

9.29.   How does the ability of a rock to withstand tension compare with its ability to withstand compression?

9.30.   The energy source of erosional processes is the sun.   Where does the energy involved in diastrophic activity come from?

9.31.   Masses of igneous rock are found to intrude the folded sedimentary and metamorphic rocks of large mountain ranges.   What does this suggest about the time sequence of the various events in the formation of these ranges?

9.32.   What geological process is chiefly responsible for the topography of a mountain range?

Answers to Supplementary Problems

9.23.   Rhyolite, andesite, basalt, obsidian.

9.24.   Basalt.

9.25.   The various fragments are cemented together by minerals precipitated from groundwater.

9.26.   A concordant pluton is intruded between older rock beds and hence lies parallel to them; sills and laccoliths are examples.   A discordant pluton cuts across older rock beds; dikes and batholiths are examples.

9.27.   Dikes and sills are usually relatively thin and hence cool fairly rapidly after being intruded.

9.28.   The intruded magma that solidifies into a pluton is very hot, and thus nearby rocks often undergo thermal metamorphism.

9.29.   Rocks are considerably stronger under compression than under tension.

9.30.   The earth's interior.

9.31.   The first phase was the deposition of sediments in a geosyncline and their hardening into rocks, then the folding and raising of the sedimentary layers, and finally the intrusion of plutons.

9.32.   Erosion.

Chapter 10

The Earth's Interior

SEISMIC WAVES

The majority of earthquakes are caused by the sudden displacement of crustal blocks along a fault, which relieves stresses that have built up over a period of time. An earthquake gives rise to waves of three kinds:

1. *Primary (or P) waves,* which are longitudinal waves that involve back-and-forth vibrations of particles of matter in the same direction as that in which the waves travel. P waves are essentially pressure waves, like sound waves. Waves that travel down a coil spring when one end is pulled out are longitudinal (Fig. 10-1).

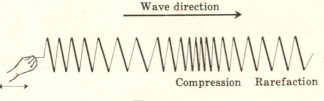

Fig. 10-1

2. *Secondary (or S) waves,* which are transverse waves that involve vibrations of particles of matter perpendicular to the direction in which the waves travel. The waves produced by shaking a stretched string are similar to S waves (Fig. 10-2). Both P and S waves are body waves that travel through the earth's interior. S waves cannot occur in a liquid, though P waves can.

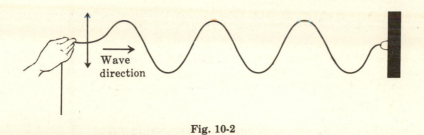

Fig. 10-2

3. *Surface (or L) waves,* which are analogous to water waves, involve orbital motions of particles of matter and are limited to the earth's surface (Fig. 4-1).

The three kinds of seismic waves travel with different velocities and so arrive at a distant observer at different times after an earthquake; P waves are the first, next S waves, and finally L waves. The velocities of P and S waves increase with depth, so that refraction causes them to travel along curved paths through the earth. In addition, both refraction and reflection occur at boundaries between regions with different physical properties. The analysis of seismic waves received at observatories around the world has led to the identification of three principal regions within the earth.

INTERIOR STRUCTURE

The earth consists of a central *core* 2160 mi (3500 km) in radius; a surrounding *mantle* 1800 mi (2900 km) thick; and a *crust* whose thickness ranges from typically 5 mi under the oceans to an average of 20 mi under the continents (Fig. 10-3).

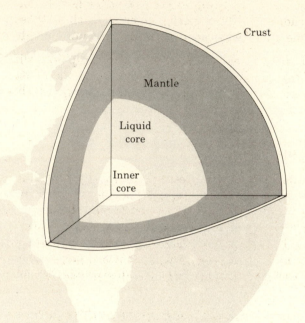

Fig. 10-3

The core, which constitutes 19% of the earth's volume, is thought likely to be composed largely of molten iron with some nickel and perhaps traces of sulfur and silicon.   The inability of S waves to travel through the core is strong evidence for its liquid nature, since transverse waves can only be transmitted by solid materials.   The inner part of the core behaves differently from the rest and is probably solid; this *inner core* has a radius of about 780 mi (1300 km).

The mantle, which constitutes 80% of the earth's volume and about 67% of its mass, is solid and is thought likely to be composed of ferromagnesian silicate minerals such as olivine, pyroxene, and garnet.   The crust consists of a global layer several miles thick of simatic (basaltic) rock with thicker layers of sialic (granitic) rock under the continents.   The crust-mantle boundary is called the *Mohorovičić discontinuity* (Fig. 10-4).

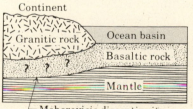

Fig. 10-4

LITHOSPHERE AND ASTHENOSPHERE

The crust and the outermost part of the mantle comprise a rigid shell of rock 50 to 100 km thick called the *lithosphere*. A layer about 100 km thick in the mantle below the lithosphere, known as the *asthenosphere,* is apparently capable of plastic flow, unlike the rigid lithosphere and the rest of the mantle. Stresses applied to the asthenosphere over a long period of time, such as the weight of a continental block or the horizontal forces exerted during continental drift, cause the astheno- sphere to flow gradually. Short-period stresses, such as those produced by an earthquake, are transmitted in the same way as in a rigid material, though with a reduced velocity.

The lithosphere in essence floats on the asthenosphere, with irregularities such as the conti- nental blocks being supported by their buoyancy in the denser material of the asthenosphere. This concept, called *isostasy,* is supported by the observation that the higher a mountain range is, the deeper are its roots.

GEOMAGNETISM

The earth's magnetic field closely resembles the field that would be produced by a giant bar magnet located near the earth's center and tilted by 11° from the axis of rotation. Since the core is molten, no such magnet can actually exist there. Instead, the field is believed to arise from coupled fluid motions and electric currents in the liquid iron of the core; a current in the form of a loop is surrounded by a magnetic field of the same form as that of a bar magnet. The geomagnetic field cannot originate in the mantle because it is a nonconductor, hence no electric currents can exist there, and because it is too hot for a ferromagnetic substance such as iron to retain its magnetism.

Figure 10-5 shows the configuration of the earth's magnetic field with the help of *lines of force.* These are imaginary lines whose direction at any point is that of the field and whose spacing varies with the strength of the field: the closer together the lines are drawn, the stronger the field in that region. The *geomagnetic poles* are points where the magnetic axis passes through the earth's surface.

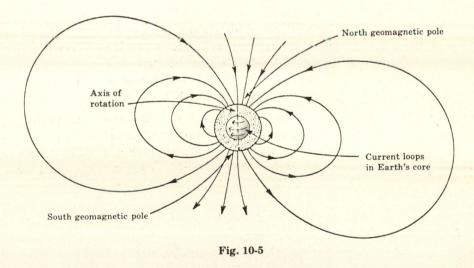

Fig. 10-5

Measurements of the magnetization of crustal rocks indicate that the geomagnetic field has often reversed its direction in the past. Such reversals seem consistent with the hypothesis that the field is due to electric currents in the core, since changes may well occur in the patterns of flow in the liquid iron there from time to time.

Solved Problems

10.1. What is the difference between the *focus* of an earthquake and its *epicenter*?

The location in the earth of the rock displacement responsible for an earthquake is its focus, and the point on the earth's surface directly above the focus is the earthquake's epicenter (Fig. 10-6).

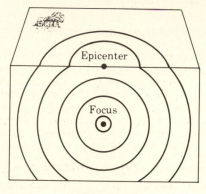

Fig. 10-6

10.2. What is a seismograph? How does it operate?

A seismograph is a device that records the vibrations produced by an earthquake. Seismographs are of two types, those that respond to vertical motions and those that respond to horizontal ones. A vertical seismograph and two horizontal ones, one each for the north-south and east-west directions, are needed at an observatory. Figure 10-7 shows the principle of operation of a vertical seismograph. The suspended mass has a very long period of oscillation, hence it remains nearly stationary in space as the scale moves up and down when seismic waves arrive. The horizontal seismograph functions in a similar way, as in Fig. 10-8, except that it records only horizontal motions relative to a steady mass.

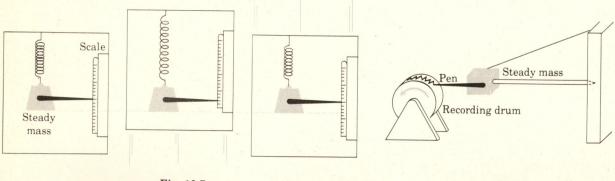

Fig. 10-7   **Fig. 10-8**

10.3. Distinguish between the *magnitude* of an earthquake and its *intensity*.

The magnitude of an earthquake is a measure of the maximum amplitude of its vibrations and is based upon seismograph records. The energy released by an earthquake can be established from its magnitude.

The intensity of an earthquake is a measure of the disturbance it produces in a particular locality, based on the destruction it causes and on the sensations experienced by people there. The intensity of an earthquake is greatest near the epicenter and decreases with distance; the magnitude of an earthquake depends only upon the energy it releases.

10.4. Describe the Richter scale of earthquake magnitude.

Each step of 1 on the Richter scale represents a change in vibrational amplitude of a factor of 10 and a change in energy release of a factor of about 30; thus an earthquake of magnitude 5 produces vibrations 10 times larger than one of magnitude 4 and evolves 30 times more energy.   An earthquake of magnitude 0 is barely capable of being detected, and the energy released, if it could be concentrated, is just about sufficient to blow up a tree stump. An inhabited area will suffer some damage if a magnitude 4.5 quake occurs nearby, and one of magnitude 6 or more may lead to significant destruction.   The energy associated with a magnitude 6 earthquake is equivalent to that of a medium-sized nuclear bomb, though its effects are different because the earthquake energy is spread out over a much wider area.   The energy released in a magnitude 8.6 earthquake, the greatest that have occurred to date, is about double the energy content of the coal and oil produced each year in the entire world; the Alaska earthquake of 1964 was of nearly this magnitude.

10.5.   **Describe the Mercalli scale of earthquake intensity.**

The Mercalli scale, in the modification in current use, has twelve degrees of earthquake intensity which are designated by roman numerals to avoid confusion with earthquake magnitude.   A disturbance of intensity I is too weak to be perceptible by most.   The vibrations of intensity III seem indoors like those caused by a truck passing nearby and may not be recognized as being due to an earthquake; they are not usually felt outdoors. VI is felt by everyone, with objects falling from shelves and trees shaking visibly, but actual damage is minor.   Structural damage to buildings begins with VII and by IX the damage is considerable and cracks appear in the ground. Destruction is total at intensity XII, with large rock masses shifted and objects thrown into the air.

10.6.   **How is it possible to tell from the seismograph record of the waves from an earthquake how far away the quake occurred?**

The various waves from an earthquake have different speeds and so require different lengths of time to go from their source to a particular location on the earth's surface.   The time interval between arrivals of each kind of wave at an observatory can be compared with the known travel-time-versus-distance data to find the distance that corresponds to this interval.

10.7.   **Figure 10-9 shows how the travel times of P, S, and L waves vary with distance (as measured on the earth's surface) from an earthquake.   (a) Find the time intervals between the arrivals of the P, S, and L waves of earthquakes that occur 6000 km from a seismograph.   (b) The S waves of an earthquake arrive at a seismograph 10 min after the P waves.   How far away did the earthquake occur?   When will the L waves arrive?**

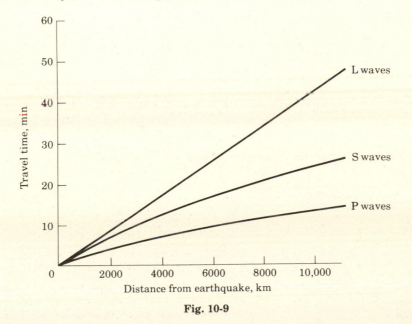

Fig. 10-9

(a)   According to Fig. 10-9, at a distance of 6000 km the P waves arrive approximately 10 min after the time of an earthquake, the S waves arrive 17 min afterward, and the L waves arrive 26 min afterward.   Hence the S waves arrive 7 min later than the P waves and the L waves arrive 9 min later than the S waves.

(b)   From the graph, a time interval of 10 min between the time-distance curves for S and P waves corresponds to a distance of roughly 8800 km.   At this distance the L waves arrive about 16 min after the S waves.

10.8. The travel times of seismic waves depend only on the distance between earthquake and observing station and do not vary around the earth for the same such distance. What does this indicate about the uniformity of the material in the earth's interior?

Because travel times depend only on distance and not on location, any variations in the material of the interior can only occur along a radius and not transversely. Thus the division of the interior into an inner core, an outer core, and a mantle, which form concentric shells, is consistent with the above observation, but a difference between the material in, say, the northern and southern hemispheres is ruled out.

10.9. Seismic S waves are never detected beyond about 7000 miles (measured along the earth's surface) from an earthquake. P waves also disappear at this distance, but reappear at distances greater than about 10,000 miles. How do these observations fit in with the hypothesis that the earth has a liquid core?

The presence of a liquid core affects both S and P waves: the S waves cannot travel through it at all, and the P waves travel in it with a different velocity than in the mantle. The latter fact means that P waves entering the core change their directions due to refraction, as shown in Fig. 10-10. As a result a "shadow zone" occurs in a band around the earth in which P waves are not found, and there is a still larger region in which S waves are absent since they are absorbed in the core. (The velocities of P and S waves vary with depth, which causes their paths to be curved due to refraction, in addition to the sharp change in direction of P waves at the core-mantle boundary.)

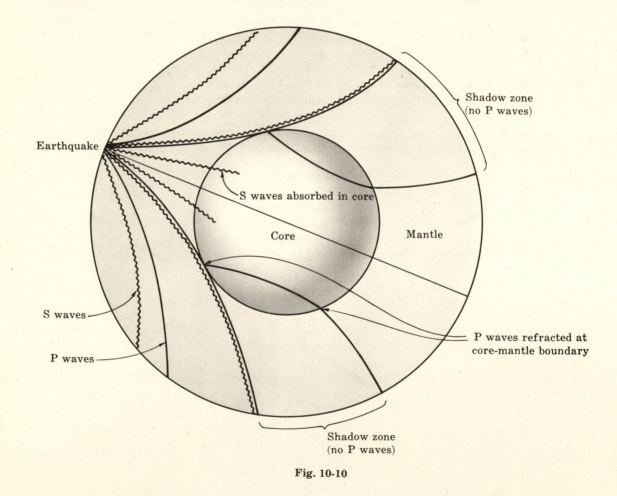

Fig. 10-10

10.10 The earth's radius is 6.4×10^6 m and its mass is 6.0×10^{24} kg. Calculate the average density of the earth as a whole and compare it with the average density of crustal rocks, 2.7 g/cm³. What does this comparison indicate about the composition of the earth's interior?

The volume of a sphere of radius r is $(4/3)\pi r^3$, hence the earth's volume is

$$V = \frac{4}{3}\pi r^3 = \frac{4}{3} \times \pi \times (6.4 \times 10^6 \text{ m})^3 = 1.1 \times 10^{21} \text{ m}^3$$

and its density is

$$d = \frac{\text{mass}}{\text{volume}} = \frac{6.0 \times 10^{24} \text{ kg}}{1.1 \times 10^{21} \text{ m}^3} = 5.5 \times 10^3 \frac{\text{kg}}{\text{m}^3} = 5.5 \frac{\text{g}}{\text{cm}^3}$$

The average density of the earth as a whole is about twice the average density of crustal rocks, which signifies that the materials of the earth's interior are much denser than those of the surface.

10.11. **What is the difference between the earth's crust and its lithosphere?**

The crust is distinguished from the mantle beneath it by a sharp difference in seismic wave velocity, which suggests a difference in the composition of the minerals involved, or in their crystal structure, or in both. The lithosphere is distinguished from the asthenosphere beneath it by a difference in their behaviors under stress: the lithosphere is rigid whereas the asthenosphere is capable of plastic flow.

10.12. **What evidence is available from seismic wave studies that supports the existence of the asthenosphere?**

The velocity of seismic waves is lower in the asthenosphere than above or below it in the mantle, which suggests that its physical properties are different. The difference is consistent with a plastic rather than a rigid character for the asthenosphere material.

10.13. **How is a plastic asthenosphere possible with a rigid lithosphere above it and a rigid mantle below it?**

The asthenosphere is plastic because its material is close to its melting point under the conditions of temperature and pressure found in that region of the mantle. Above the asthenosphere the temperature is too low, and below it the pressure is too high, for the material of the mantle to be plastic.

10.14. **List some of the considerations that have led to the belief that the material of the mantle is similar to such ultramafic (mainly composed of ferromagnesian minerals) igneous rocks as eclogite and peridotite.**

(1) Only dense rocks such as these transmit seismic waves in the manner exhibited by the mantle. (2) These rocks are similar in composition to the basalt of the lower crust but are sufficiently different to be able to give rise to the observed Mohorovičić discontinuity in seismic wave transmission between crust and mantle. (3) Most volcanic magmas come from the upper mantle, and their composition is consistent with an origin in eclogite or peridotite. (4) Diamonds can form only under conditions of high temperature and pressure, such as are found in the mantle but not in the crust, and eclogite and peridotite are common in diamond-bearing rock formations which must have originated in the mantle. (5) Stony meteorites consist chiefly of the minerals olivine and pyroxene, just as these rocks do, and it is an attractive notion that both the mantle and stony meteorites had the same origin in the early solar system.

10.15. **(a) Why is it considered likely that the earth's outer core is liquid? (b) Why is the liquid thought to be largely iron? (c) Why is nickel believed to be present as well?**

(a) Transverse waves cannot propagate through a liquid, and it is observed that seismic S waves, which are transverse, are unable to pass through the core although P waves, which are pressure waves, are able to. In support of the idea that the core is a liquid is the observation that the earth's magnetic field, which originates in its interior, fluctuates in both magnitude and direction, which is hard to explain if the interior is solid but easy to explain if some of the interior is an electrically conducting liquid.

(b) Iron is a fairly abundant element in the universe; its density is just about right for an iron core to account for the total mass of the earth, given the ferromagnesian silicate composition of the mantle; and iron is a good conductor of electricity, which is necessary in order to explain the origin of the earth's magnetic field.

(c) Meteorites that contain iron always contain a small proportion of nickel as well, which suggests that these two metals occur together in the solar system.

10.16. What evidence is there in favor of the idea that the earth's interior is very hot? What temperatures are believed to occur there?

Three observations that support the notion of high interior temperatures are:

(1)   Measurements made in mines and wells indicate that temperature increases with depth.

(2)   Molten rock from the interior emerges from volcanoes.

(3)   The outer core is liquid, which means it must be at a high temperature.

The present temperature distribution within the earth is believed to increase fairly rapidly in the mantle from less than 1000 °C at its top to perhaps 3000 °C at the core boundary. The rise is slower in the core, and the temperature at the center of the earth is estimated to be in the neighborhood of 4200 °C, though this figure is far from being certain.

10.17. What origins are likely for the high temperatures believed to exist in the earth's interior?

Radioactive uranium, thorium, and potassium isotopes inside the earth evolve considerable energy, which is manifested there as heat.   In radioactive decay, certain unstable atomic nuclei give off energy as they are spontaneously transformed into nuclei of other kinds; the subject is discussed in Chapter 12.   The earth is thought to have come into being 4.5 billion years ago as a cold aggregate of smaller bodies of metallic iron and silicate minerals that had been circling the sun.   Heat due to radioactivity accumulated in the interior of the infant earth and in time caused it to melt.   The influence of gravity then caused the iron to migrate inward to form the core while the lighter silicates rose to form the mantle.   Today most of the earth's radioactivity is concentrated in the crust and upper mantle, where the heat it produces escapes through the surface and cannot collect to remelt the interior.

Part of the earth's heat may have a different origin.   When the earth was formed, the initial matter came together and contracted under the influence of gravity into a tightly packed aggregate.   In this process the initial gravitational potential energy became heat, and some of this heat may still be present since the mantle is a poor conductor.

10.18. The earth's interior consists of a liquid layer (the outer core) sandwiched between the solid inner core and the solid mantle.   How can such a curious structure have come into being?

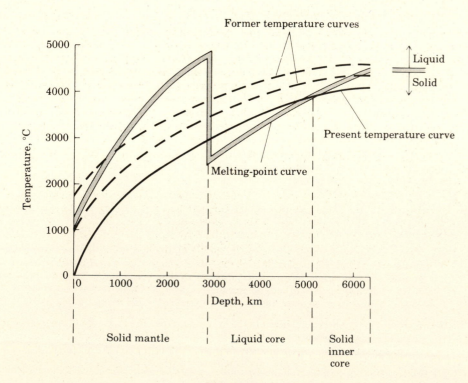

Fig. 10-11

The key to this problem is the melting point versus depth curve of the earth's materials shown in Fig. 10-11, which was obtained from theoretical considerations. At first, with increasing depth the temperature at which melting occurs increases owing to the greater pressures. At the boundary between the mantle and the core there is a sharp drop, corresponding to the different composition of the core, followed by another rise, again due to the mounting pressure.

Let us now consider the cooling of the earth from an initial molten state. The variation of temperature with depth is shown by the successive thin lines, each corresponding to a different period in the earth's history. As the earth cooled, the material that would have solidified first is that whose melting point is reached first. According to the diagram, this would have been the innermost part of the mantle. As time went on, the mantle continued to solidify outward from its boundary with the core. Somewhat later the center of the earth became cooled below its melting point, and the solid inner core thus began to come into being. Today the temperature distribution is presumably similar to that represented by the lowest temperature-depth curve, with the entire mantle and the central part of the core, both solid and separated from each other by molten nickel-iron.

10.19. Does a compass needle point toward true north everywhere on the earth's surface?

A compass needle at a certain location aligns itself with the direction of the geomagnetic field there. If the field were perfectly regular, that is, if the pattern of its lines of force were exactly that of a bar magnet or a current loop, the compass needle would point to the geomagnetic north pole everywhere. However, because the axis of the geomagnetic field is tilted by 11° from the earth's geographic axis and is centered several hundred km from the earth's center, the geomagnetic poles are displaced from the geographic poles. In addition, the geomagnetic field is somewhat irregular, so that in many parts of the earth a compass needle does not even point toward the geomagnetic north pole. In the continental United States, a compass needle will point as much as 20° east or west of true north, depending upon the location.

10.20. At a given place, are the direction and strength of the earth's magnetic field constant?

The geomagnetic field varies in both direction and strength. In the past century, for instance, the average field strength has decreased by about 6%. Local changes in direction of a degree or so per year are common. The most striking changes have been reversals in the field direction, so that the geomagnetic north and south poles become exchanged. Evidence from magnetized rocks that have hardened at different times suggests that in the past 76 million years 171 field reversals have occurred. Such reversals apparently involve a decrease in the field strength to zero over a period of thousands of years, followed by the buildup of the field with the opposite polarity.

10.21. What kinds of rocks exhibit fossil magnetism?

Igneous, metamorphic, and sedimentary rocks can all exhibit fossil magnetism and thus indicate the direction of the geomagnetic field at the time they were formed. All magnetic materials lose their magnetic properties above a certain temperature (called the *Curie point*) which varies with the nature of the material; the Curie point is less than 800 °C for nearly all common substances. When an igneous rock containing magnetizable minerals cools from a molten state, it is magnetized parallel to the local geomagnetic field. A metamorphic rock will behave the same way if it has been heated above the Curie point of its minerals; melting is not necessary. Magnetic mineral grains will tend to be aligned parallel to the local geomagnetic field when they are deposited as sediments, so it is also possible for sedimentary rocks to exhibit fossil magnetism.

10.22. What is the source of energy of the dynamo mechanism responsible for the earth's magnetic field?

The geomagnetic field arises from coupled fluid motions and electric currents in the outer core. The energy source is heat from the inner core that causes convection in the outer core which is influenced by the earth's rotation to give the flow patterns involved in the production of the magnetic field. The heat of the inner core is thought to be heat of fusion which is released by the continuing crystallization of the molten nickel-iron of the outer core as the inner core grows in size. Radioactivity in the inner core is another source of heat, but it is probably too small to account for the required energy.

Supplementary Problems

10.23. How many seismological observatories must record the waves sent out by an earthquake in order to locate its epicenter?

10.24. What are the similarities between seismic P and S waves? What are the differences?

10.25. Do P and S waves travel at constant velocities through the earth?

10.26. The S waves of an earthquake arrive at a seismograph station 5 min after the P waves. How far away did the earthquake occur?

10.27. An earthquake occurs 10,000 km from a seismograph station. How many minutes later do the P, S, and L waves arrive there?

10.28. At what depths do earthquake foci occur?

10.29. Why cannot earthquakes occur at great depths in the earth?

10.30. Where is the earth's crust thinnest? Where is it thickest?

10.31. How does the radius of the earth's core compare with the radius of the earth as a whole? How does the mass of the core compare with the mass of the earth as a whole?

10.32 Why is it believed that the radioactive materials in the earth are concentrated in the crust and upper mantle?

10.33. The earth's magnetic field is very nearly the same as that which would be produced by a loop of electric current located in the interior of the earth. Assuming that the field is exactly the same as that of a current loop, what is the simplest procedure you could use to find the locations of the geomagnetic poles, which are those points where the axis of the loop intersects the earth's surface?

10.34. Why is it unlikely that the earth's magnetic field originates in a huge bar magnet located in its interior?

Answers to Supplementary Problems

10.23. An observatory can establish the distance of the earthquake from its own location, which means that the epicenter must be located somewhere on a circle of this radius drawn on a globe with the observatory as the center. The circles thus drawn from two observatories will in general intersect at two places, so a third circle is needed to determine a unique location for the epicenter. Hence three observatories must record the waves from an earthquake in order to locate it.

10.24. P and S waves both can travel through a solid medium, but only P waves can travel through a liquid. P waves are longitudinal and, like sound waves, consist of pressure fluctuations; S waves are transverse and are analogous to waves in a stretched string. P waves are faster than S waves in the same medium.

10.25. The velocities of both P and S waves increase with depth in the mantle. The velocity of P waves drops sharply when they enter the core, and then increase with depth toward the earth's center.

10.26. Approximately 4000 km away.

10.27. The P waves will arrive after about 13 min, the S waves after about 24 min, and the L waves after about 43 min.

10.28. Earthquake foci occur at all depths from the surface down to about 700 km. Most earthquakes have focal depths of less than 75 km; only about 3% have focal depths below 300 km.

10.29.   The greater the depth, the greater the pressure and hence the greater the force needed to produce the sudden movement along a fault that causes an earthquake.   At extreme depths it is impossible for the force needed to be generated.

10.30.   The crust is thinnest under the oceans and thickest under the continents.

10.31.   The core's radius is about 55% of the radius of the earth; the core's mass is about one-third the mass of the earth.

10.32.   So much energy is given off by the earth's radioactivity that the whole interior would be molten if it were not concentrated in the outer part of the earth.

10.33.   A magnetized needle suspended so it can move freely will point vertically downward at the geomagnetic poles since the direction of the magnetic field at these points is along the axis of the loop.

10.34.   Ferromagnetic materials lose their magnetic properties at high temperatures, and sufficiently high temperatures exist throughout all of the earth's interior except near the surface of the crust.   Also, both the direction and strength of the field are observed to vary, and in fact the field has reversed its direction many times in the past, which cannot be reconciled with the notion of a permanent magnet in the interior.

Chapter 11

Continental Drift

THE OCEAN FLOORS

The earth's crust is subject not only to vertical changes, with entire regions being thrust upward and other regions subsiding, but also to horizontal changes, with the continents continually shifting their positions around the earth. The evidence for the latter events is recent but conclusive. Some of the major findings concern the ocean floors:

1. The ocean floors are relatively recent in origin; the oldest sediments date back only about 135 million years, in contrast to continental rocks which date back as much as 4000 million years. Many parts of the ocean floor are much younger still, so that about one-third of the earth's surface has come into existence in 1.5% of the earth's history.

2. A worldwide system of narrow *ridges* and somewhat broader *rises* runs across the oceans. An example is the Mid-Atlantic Ridge, which virtually bisects the Atlantic Ocean from north to south; Iceland, the Azores, and Ascension Island are some of the higher peaks in this ridge.

3. The direction of magnetization of ocean-floor rocks is the same along strips parallel to the midocean ridges, but the direction is reversed from strip to strip going away from a ridge on either side.

4. A system of *trenches* several km deep occurs around the rim of the Pacific Ocean; it coincides with the belt in which most current earthquakes and volcanoes occur. The trenches have *island arcs* on their landward sides that consist of volcanic mountains projecting above sea level.

PLATE TECTONICS

The preceding observations are accounted for by the theory of *plate tectonics*. According to this theory, the lithosphere is divided into seven very large *plates* plus a number of smaller ones, all of which float on the plastic asthenosphere (Fig. 11-1). Three kinds of events may occur at a boundary between adjacent plates, as in Fig. 11-2.

1. **Plate creation.** The plates move apart at several cm per year and molten rock rises to form new ocean floor on either side. A midocean ridge occurs at this boundary and consists of the latest rock to be deposited. As the new rock hardens, it is magnetized in the same direction as the geomagnetic field at the time; since this field reverses its direction several times per million years on the average, the result will be strips with alternate magnetization on both sides of the ridge, as observed.

2. **Plate destruction.** One plate slides underneath the other and melts when it reaches the mantle. An oceanic trench is produced at such a *subduction zone*. Volcanoes and island arcs are formed where the less dense components of the descending plate margin melt and rise to the surface. When continental blocks on adjacent plates are pressed together in a subduction zone, they are too light relative to the underlying material for either to be forced under, and instead they buckle to produce a mountain range.

3. **Plate motion.** At some boundaries the adjacent plates are simply sliding past each other without colliding or moving apart. These boundaries, where only horizontal motion occurs, are called *fracture zones*; earthquakes are frequent along them because of friction

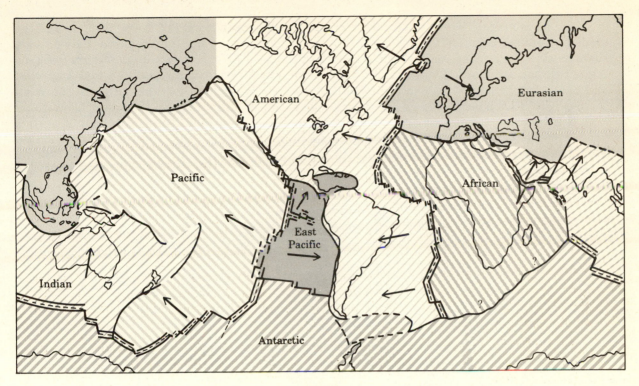

Fig. 11-1

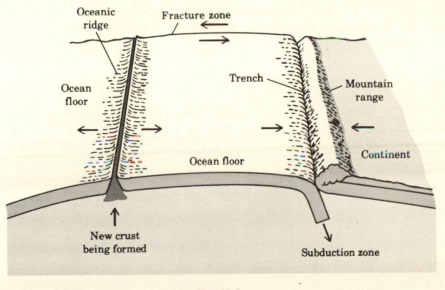

Fig. 11-2

between the plate margins. An ideal plate would thus have one edge growing at a ridge, the opposite edge disappearing into the mantle, and the other two edges sliding past the edges of adjoining plates.

CONTINENTAL DRIFT

As the various lithospheric plates shift around the earth, growing at some margins and being destroyed at others, the continents shift with them. Geological, magnetic, climatic, and biological evidence have made possible the reconstruction of past arrangements of the continents and the prediction of future ones.   It seems likely that about 200 million years ago there was only a single supercontinent, Pangaea, and a single ocean, Panthalassa.   After some millions of years Pangaea began to break apart into Gondwanaland (South America, Africa, Antarctica, India, and Australia) and Laurasia (Eurasia, Greenland, and North America); the Tethys Sea came into being between them (Fig. 11-3).   South America and Africa then broke off as a unit from the rest of Gondwanaland, and later they separated as the South Atlantic Ocean came into being.   By about 65 million years ago the Atlantic Ocean had completed its extension northward, Australia had separated from Antarctica, and India had begun to drift toward Asia.   About half the present ocean floor—a third of the earth's crust—has come into being since then and the same amount of old crust has been destroyed; all in 1.5% of the earth's history.

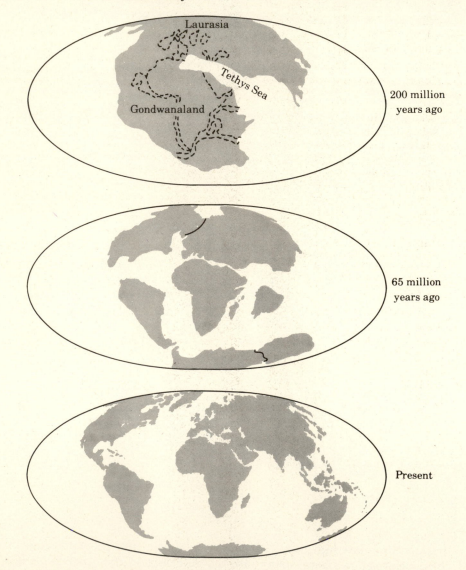

Fig. 11-3

Some tens of millions of years from now the Atlantic Ocean will be wider than it is today and the Pacific will be narrower.   Australia will have moved north.   California will have been detached from the rest of North America, the Arabian peninsula will be attached to Asia, the Mediterranean will be much smaller, and East Africa will have broken away from the rest of Africa.

Solved Problems

11.1. **Why are the ocean floors so much younger than the continents?**

Owing to their low density and consequent buoyancy, the continental blocks are not forced down into the mantle in subduction zones but remain as permanent features of the lithospheric plates they are part of. The ocean floors, on the other hand, are continually being destroyed in such zones, as new ocean floors are deposited at midocean ridges.

11.2. **When continental drift was first proposed half a century ago, it was assumed that the continents move through soft ocean floors. Why is this hypothesis no longer considered valid? How does continental drift actually occur?**

The ocean floors are extremely rigid, so it is not possible for the continental blocks to move through them. Instead, the continents are each part of a lithospheric plate whose motion is accomplished by the destruction of the plate at one margin and the formation of new plate at the opposite margin.

11.3. **How can observations of the magnetization of rocks provide information on continental drift?**

When a sediment that includes iron-containing minerals is deposited, or when a magma with iron-containing minerals hardens, it becomes weakly magnetized by the earth's magnetic field. The direction of magnetization is the same as that of the lines of force of the earth's field at the time the rock was formed, hence it indicates the latitude at which the rock was then located and the positions of the magnetic poles then. Under the assumption that the magnetic and geographic axes have never been far apart—which is supported both experimentally and theoretically—it is possible to correlate such paleomagnetic evidence from rocks of different ages and from different locations into a picture of continental drift that agrees with the picture derived from geological findings.

11.4. **Distinguish between *transcurrent* and *transform* faults.**

A transcurrent fault is a strike-slip fault along whose entire length relative displacement of the two sides occurs. Figure 11-4a shows a left-lateral transcurrent fault; the ridge crests on both sides of the fault are moving apart.

A transform fault is a strike-slip fault along which relative motion occurs only *between* offset ridge crests on both sides of the fault. New crust that spreads out from the ridge crests produces motion along the fault, but the crests themselves keep their original separation. Figure 11-4b shows a right-lateral transform fault.

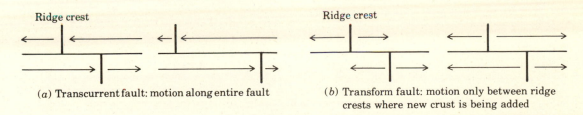

(a) Transcurrent fault: motion along entire fault   (b) Transform fault: motion only between ridge
                                                            crests where new crust is being added

Fig. 11-4

11.5. **There is strong evidence that today's continents were once parts of a single supercontinent, Pangaea. What can be said about the likelihood that continental drift occurred before Pangaea came into existence?**

The motions of lithospheric plates are on such a huge scale that it is hard to believe they suddenly began only 200 million years ago, which is relatively recent in the geological history of the earth. Hence it is likely that continental drift was taking place even before Pangaea was formed, and in fact there is some evidence (such as the existence of the Ural Mountains) that Pangaea was the result of the coming together of three earlier continents, Gondwanaland, Asia east of the Urals, and a land mass consisting of North America, Greenland, and Europe.

11.6. **What kind of biological evidence supports the notion that all the continents were once part of a single supercontinent?**

Until perhaps 180 million years ago fossils indicate that living things of the same kinds occurred everywhere that suitable habitats existed, whereas since that time many plants and animals have evolved differently in different continents.

11.7. Give examples of climatological findings that support the concept of continental drift.

(1) Glacial deposits and evidence of glacial erosion are found in tropical regions of South America, Africa, and India, which suggests that these regions were once much farther south.

(2) Coal is formed from plant debris that accumulates in swamps, and its presence in Antarctica suggests that this continent was once much farther north.

(3) *Evaporites* are sedimentary rocks composed of minerals such as halite, gypsum, and anhydrite that have precipitated from solution in water. Evaporites form in arid climates such as occur in the "horse latitudes," which are high-pressure regions centered on latitudes 30° N and 30° S (see Problem 3.10). Deposits of evaporites several hundred million years old have been found at much higher latitudes, which suggests that these regions must then have been closer to the equator.

(4) Coral reefs grow only within about 30° of latitude on either side of the equator. Traces of ancient coral reefs are found well to the north of this belt, which suggests that they developed at a time when these regions were closer to the equator than they are today.

11.8. What is the *Gondwana succession?*

Gondwana is a district of India in which occurs a distinctive sequence of beds, including tillite (consolidated glacial debris) and coal deposits, laid down from about 400 million to 200 million years ago. The same "Gondwana succession" of beds is found in South America, Africa, Australia, and Antarctica. An important feature of the Gondwana succession is the presence of certain plant fossils which are the same in beds of the same age throughout the southern hemisphere. Thus the geological and biological records point strongly to the existence of a single land mass, Gondwanaland, in the southern hemisphere in the distant past.

11.9. (*a*) What mountain ranges of today were once part of the Tethys Sea? (*b*) What kind of evidence would indicate that the region where these mountains are present was once below sea level?

(*a*) The Pyrenees, Alps, and Caucasus of Europe; the Atlas Mountains of North Africa; and the Himalayas of Asia.

(*b*) Thick deposits of sedimentary rocks; fossils of sea creatures.

11.10. The east coast of South America is a good fit against the west coast of Africa. What sort of evidence would you look for to confirm that the two continents had once been part of the same land mass?

If South America and Africa were once joined together, there should be similar geological formations and fossils of the same kinds at corresponding locations along their respective east and west coasts. This is indeed found for material deposited up to about 100 million years ago, which is when these continents must have begun to separate.

11.11. Madagascar is about 2700 million years old whereas Providence Island, which is not far away from it, is 36 million years old. What do these ages suggest about the origins of these islands?

Madagascar must have broken off from the African continent, whereas Providence Island must have a volcanic origin.

11.12. North America, Greenland, and Eurasia fit together reasonably well in reconstructing Laurasia, but there is no space available for Iceland. Why is the omission of Iceland from Laurasia acceptable?

Iceland is less than 70 million years old, much younger than North America, Greenland, and Eurasia, and was formed after the breakup of Laurasia from magma rising through the rift in the Mid-Atlantic Ridge.

11.13. How does the origin of the Himalayas differ from that of the oceanic mountains that constitute the Mid-Atlantic Ridge?

 The Himalayas were thrust upward by the collision of the Indian plate and the Eurasian plate. The Mid-Atlantic Ridge was formed by the upwelling of molten rock.

11.14. How would you expect the ages of the South Pacific islands far away from the East Pacific Rise to compare with those near this rise?

 The East Pacific Rise marks the rift through which magma wells to the surface to form new ocean floor, hence islands far from the rise were formed earlier than those near the rise.

11.15. The distance between the continental shelves of the east coast of Greenland and the west coast of Norway is about 1300 km. If Greenland separated from Norway 65 million years ago and their respective plates have been moving apart ever since at the same rate, find the average velocity of each plate.

 Each plate must have moved half of 1300 km, or 650 km, in 65×10^6 years, and so, since $1 \text{ km} = 10^5$ cm, its velocity is

$$v = \frac{\text{distance}}{\text{time}} = \frac{650 \text{ km} \times 10^5 \text{ cm/km}}{65 \times 10^6 \text{ yr}} = 1 \text{ cm/yr}$$

11.16. List the most plausible suggestions that have been made for the mechanism by which lithospheric plates are moved across the earth's surface. Are any of these ideas widely accepted?

 (1) Convection currents in the asthenosphere.

 (2) The sinking of the edge of a lithospheric plate in a subduction zone pulls the rest of the plate across the earth's surface.

 (3) The upwelling magma in a midocean ridge pushes the adjacent plates apart.

 (4) The elevated material in a midocean ridge forces the adjacent plates apart by virtue of its weight, much as in the case of a man standing with his feet in two adjacent rowboats.

 Each of these ideas is open to serious objections and none is at present widely accepted by geologists. It is possible that several mechanisms contribute to plate motion.

Supplementary Problems

11.17. Which of today's continents were once part of Laurasia? Of Gondwanaland?

11.18. Is the rate of ocean-floor spreading the same everywhere?

11.19. Why is it believed that the plate material that descends into the earth in a subduction zone may not melt completely until it reaches depths of as much as 600–700 km?

11.20. Why is there a mountain range on the western edge of South America but not one on its eastern edge?

11.21. The San Andreas Fault in California is a strike-slip fault that lies along the boundary between the Pacific and American plates. What does this indicate about the nature of the boundary?

11.22. If the entire mantle were to become as rigid as the lithosphere, so that vertical motions due to isostasy and to plate tectonics would no longer occur, would erosion eventually lead to a perfectly smooth, water-covered earth?

11.23. The oldest sediments found on the floor of the South Atlantic Ocean 1300 km west of the axis of the Mid-Atlantic Ridge were deposited about 70 million years ago. What rate of plate movement does this finding suggest?

Answers to Supplementary Problems

11.17.   Laurasia: North America, Greenland, Eurasia (except India).   Gondwanaland: South America, Africa, Antarctica, Australia, India.

11.18.   Observed rates vary from less than 1 cm/yr to over 10 cm/yr.

11.19.   The deepest earthquake foci at subduction zones occur at such depths.

11.20.   The boundary of the western Atlantic plate is along the western edge of South America, and the Andes are the result of the collision between this plate, which is drifting westward, and the eastern Pacific plate, which is drifting eastward.   The eastern edge of South America is in the interior of the western Atlantic plate and hence is geologically stable.

11.21.   The San Andreas Fault is part of a fracture zone; the Pacific plate is moving northwestward relative to the American plate, and this fracture zone forms the boundary between them.

11.22.   Not necessarily.   Local heating could lead to volcanic activity and hence to the formation of mountains, and local subsidence could occur due to the solvent action of groundwater.

11.23.   1.9 cm/yr.

Chapter 12

Earth History

RELATIVE TIME

Before the development of radioactive methods in this century, the geological events that have shaped the earth's surface could only be placed in historical order. The principles of historical geology, the chief of which are listed below, permit a relative time scale to be determined for many such events.

1. The geological processes that occurred in the past are the same as the ones that are occurring today, although not necessarily in the same places or to the same extent. This is the *principle of uniform change,* or *uniformitarianism.*

2. In a sequence of sedimentary rocks, the lowest bed is the oldest and the highest bed is the youngest. The beds were originally deposited in horizontal layers.

3. Folding and faulting occurred later that the youngest bed affected.

4. An igneous rock is younger than the youngest bed it intrudes.

FOSSILS

Fossils are the remains or other traces of living things of the past that are found in rocks. Organisms have evolved continuously from simple to complex forms and have also changed in response to environmental changes, and the fossil record mirrors this progression. Fossils are accordingly useful in historical geology. For instance, a series of rock beds can be arranged in the sequence of their formation on the basis of the fossils they contain, which may not be possible from purely geological considerations. In addition, beds deposited at the same time but in different places can be correlated from the presence in them of fossils of the same kinds. The type of fossil found in a particular bed also reveals something about the local environment at the time the organism was alive: whether the region was dry land or was covered by fresh or salt water, whether the climate was warm or cold, and so forth.

RADIOACTIVITY

The atoms of a particular element all have the same number of protons in their nuclei, which is the *atomic number* of the element. However, the number of neutrons in the nuclei may differ. For instance, all chlorine nuclei contain 17 protons, but about three-quarters of them contain 18 neutrons and one-quarter contain 20 neutrons. The varieties of an element with different nuclear compositions are called its *isotopes.* An isotope is usually identified by its *mass number,* which is the total number of protons and neutrons in the nucleus of one of its atoms; thus the isotope of chlorine corresponding to a nucleus with 18 neutrons is referred to as chlorine 35 (or ^{35}Cl).

Certain isotopes are unstable and undergo *radioactive decay* into more stable ones. Four common types of radioactive decay are:

1. *Alpha decay,* in which a helium nucleus that consists of two protons and two neutrons is emitted. Alpha decay occurs in nuclei too large to be stable.

2. *Beta decay,* in which an electron is emitted when one of the neutrons in a nucleus spontaneously turns into a proton. Beta decay occurs in nuclei in which the neutron-proton ratio is too large for stability.

3. *Electron capture,* in which one of the inner electrons in an atom is absorbed by one of the protons in its nucleus to form a neutron. Electron capture occurs in nuclei in which the neutron-proton ratio is too small for stability.

4. *Gamma decay,* in which a gamma ray (an electromagnetic wave of shorter wavelength than those of X-rays) is emitted by a nucleus with excess energy, as is often the case after one of the other types of decay occurs. Gamma decay does not change the nature of a nucleus.

RADIOACTIVE DATING

A nucleus subject to radioactive decay always has a certain definite probability of decay during any time interval. The *half-life* of a radioisotope is the time required for half of any initial quantity of it to decay. If an isotope has a half-life of, say, 2 hr and we start with 100 g of it, after 2 hr, 50 g will be left undecayed; after 4 hr, 25 g will be left undecayed; after 6 hr, 12.5 g will be left undecayed; and so on. Figure 12-1 illustrates the decay of potassium 40 (^{40}K) into argon 40 (^{40}Ar) by electron capture, which has a half-life of 1.3×10^9 (1.3 billion) years.

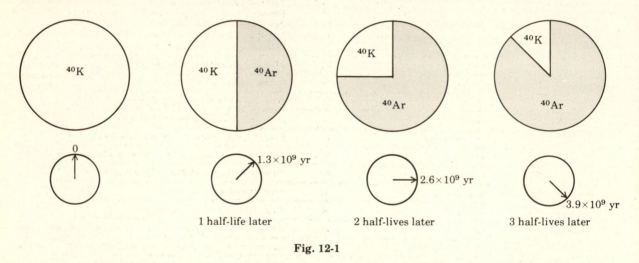

Fig. 12-1

The half-life of a radioisotope is a characteristic physical property which never changes. For this reason methods based on radioactive decay make it possible to establish the ages of many rocks on an absolute rather than a relative time scale. The ratio between the amounts of a certain nuclide and its stable daughter product in a sample therefore indicates the age of the sample; the greater the proportion of the daughter product, the older the sample.

Four radioactive nuclides found in common minerals are especially useful in dating igneous and metamorphic rocks:

Parent Radioisotope	Stable Daughter Isotope	Half-Life
Potassium 40	Argon 40	1.3×10^9 years
Rubidium 87	Strontium 87	47×10^9 years
Uranium 235	Lead 207	0.7×10^9 years
Uranium 238	Lead 206	4.5×10^9 years

Potassium 40 and rubidium 87 each decay to a stable daughter in a single step, but both uranium isotopes undergo several successive decays before becoming stable lead isotopes. The carbon 14 method (see Problem 12.12) can be used for dating sedimentary deposits that contain fossil carbon up to an age of about 40,000 years.

GEOCHRONOLOGY

Although the evolution of living things is a continuous process, the fossil record shows that three especially marked changes in the patterns of plant and animal life have taken place in the past. These times of change divide the most recent 570 million years of the earth's history into three *eras*: *Paleozoic* ("ancient life"), *Mesozoic* ("Intermediate life"), and *Cenozoic* ("recent life"). The nearly 4 billion years from the earth's formation to the start of the Paleozoic era are lumped together into *Precambrian time*. Figure 12-2 shows the currently-accepted division of the eras into periods; the various periods are further subdivided into epochs, of which only those of the Cenozoic era are shown.

Millions of years before the present	Era	Period	Epoch	Duration in millions of years	The biological record	
65	Cenozoic	Quaternary	Recent	0.01	Man becomes dominant	
225			Pleistocene	2.5	Rise of man; large mammals abundant	
570		Tertiary	Pliocene	4.5	Flowering plants abundant	Age of Mammals
			Miocene	19	Grasses abundant; rapid spread of grazing mammals	
			Oligocene	12	Apes and elephants appear	
			Eocene	16	Primitive horses, camels, rhinoceroses	
			Paleocene	11	First primates	
	Mesozoic	Cretaceous		71	First flowering plants; dinosaurs die out	Age of Reptiles
		Jurassic		54	First birds; dinosaurs at their peak	
		Triassic		35	Dinosaurs and first mammals appear	
	Paleozoic	Permian		55	Rise of reptiles; large insects abundant	
		Pennsylvanian	Carboniferous	45	Large nonflowering plants in enormous swamps	
		Mississippian		20	Large amphibians; extensive forests; sharks abundant	
		Devonian		50	First forests and amphibians; fish abundant	
4000	Oldest rocks	Silurian		35	First land plants and coral reefs	
		Ordovician		70	First vertebrates (fish) appear	
		Cambrian		70	Marine shelled invertebrates (earliest abundant fossils)	
	Precambrian time	Late Precambrian			Marine invertebrates, mainly without shells	
4500		Early Precambrian			Marine algae (primitive one-celled plants)	

Fig. 12-2

Solved Problems

12.1. An *angular unconformity* is an irregular surface that separates tilted lower rock strata from horizontal upper ones. How does such an unconformity originate?

An angular unconformity is a buried surface of erosion that involves at least four events: (1) deposition of the oldest strata; (2) diastrophic movement that raises and tilts the existing strata; (3) erosion of the elevated strata to produce an irregular surface that cuts across their exposed edges; (4) a new period of deposition that buries the eroded surface.

12.2. In Fig. 12-3, beds *A* to *F* consist of sedimentary rocks formed from marine deposits and *G* and *H* are granite. What sequence of events must have occurred in this region?

(1) Deposition of beds *E* and *F* when the region was below sea level.

(2) Diastrophic movement that produced the fault *JJ'* and the folds in beds *E* and *F*.

(3) Deposition of bed *D*.

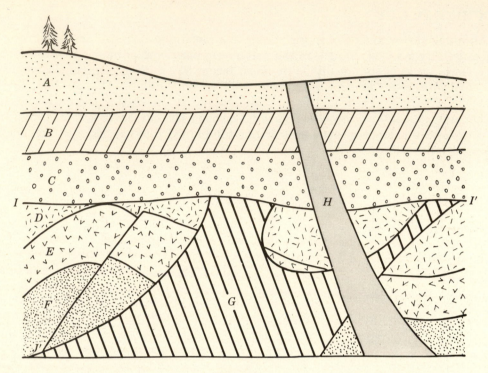

Fig. 12-3

(4) Intrusion of the granite pluton *G*.

(5) Erosion that produced the irregular surface *II'*, which is an unconformity. The region must have been elevated above sea level for this erosion to have occurred.

(6) The region subsided below sea level and beds *A, B,* and *C* were deposited.

(7) Intrusion of the granite pluton *H*.

(8) Re-elevation of the region above sea level and the renewed erosion of the surface.

12.3. **What is the biological basis for the division of geological time into eras and periods?**

The fossil record shows that there were a number of occasions when animal and plant life became sharply reduced in both number and variety, to be followed in each case by the rapid evolution of new types. The intervals of extinction are used to divide geological history into periods; during a typical period there is an expansion of living things followed by a time in which biological change is more gradual, then an interval of extinction ends the period. The division into eras is based on exceptionally marked, worldwide extinctions and subsequent expansions of plant and animal life.

12.4. **The earth's history is sometimes divided into two *eons, Cryptozoic* ("hidden life") and *Phanerozoic* ("visible life"), with the first corresponding to Precambrian time and the second extending from the beginning of the Paleozoic era to the present day. What is the reason for this division?**

Abundant fossils exist from the Phanerozoic eon, which permit tracing the evolution of living things during this span of time. Few fossils exist from the Cryptozoic eon, making it difficult to determine the forms of life that were present then and how they developed.

12.5. **List some of the various kinds of fossils.**

(1) Actual plant or animal tissues, usually of a hard nature such as teeth, bones, hair, and shells. Entire insects have been found preserved in amber.

(2)   Plant tissues that have become coal through partial decay but which retain their original forms.

(3)   Tissues that have been replaced by a mineral (such as silica) from groundwater; petrified wood is an example. Sometimes a porous tissue such as bone will have its pore spaces filled with a deposited mineral.

(4)   Impressions that remain in a rock of plant or animal structures that have themselves disappeared.

(5)   Footprints, worm holes, or other cavities produced by animals in soft ground that have later filled with a different material and so can be distinguished today.

12.6.   Why are most fossils found in beds that were once the floors of shallow seas?

Plant and animal life is abundant in such seas, and dead organisms sink to the bottom where they are soon buried in sediments that protect them from decay. Land organisms rarely leave fossils unless they fall into a swamp or lake, because their remains are subject to chemical and bacterial decay and the attacks of scavengers.

12.7.   Why are fossils still useful in dating rock formations despite the development of radioactive methods?

Fossils are found in sedimentary rocks, and radioactive dating is not generally useful in such rocks. Dating by means of fossils is usually much easier than radioactive dating, which requires elaborate apparatus. Since the geological periods that correspond to specific fossil types are well established, fossil dating is reasonably accurate.

12.8.   What are the two basic conditions that must be met by a radioactive nuclide in order that it be useful in dating a particular kind of rock?

The nuclide must occur in at least one of the minerals found in the rock, and it must have a half-life that is roughly comparable with the age of the rock (within a factor of 10 to 100, depending upon the details of the situation).

12.9.   The half-life of radium 226 is 1600 years. How long will it take 15/16 of a given sample of radium 226 to decay?

One-sixteenth is left, and since $1/2 \times 1/2 \times 1/2 \times 1/2 = 1/16$, this means four half-lives or 6400 years.

12.10.   Why is the potassium-argon method more generally useful than the other radiometric methods?

The half-life of rubidium 87 is 47 billion years, so the rubidium-strontium method can only be used to date extremely old rocks. Potassium 40 has the more suitable half-life of 1.3 billion years and is a much more widespread constituent of minerals than uranium. Such common minerals as the micas, the feldspars, and hornblende all contain sufficient potassium to permit their dating by the potassium-argon method.

12.11.   Why is it difficult to date clastic sedimentary rocks by radiometric methods?

A clastic sedimentary rock consists of fragments of other rocks that have become cemented together. The parent rocks may have been of very different ages since erosional debris is commonly transported for some distance from its origin to the place of deposition. Since the age of a sedimentary rock refers to the time it became lithified, the only relationship between the age of the rock and the ages of the fragments of which it is composed is that the rock is younger than the fragments; but it is seldom possible to say how much younger. Only in a few cases does the cementing material contain sufficient potassium to permit its dating by the potassium-argon method.

12.12.   The carbon isotope ^{14}C (called "radiocarbon") is beta-radioactive with a half-life of 5600 years. Radiocarbon is produced in the earth's atmosphere by the action of cosmic rays (described in Chapter 15) on nitrogen atoms, and the carbon dioxide of the atmosphere contains a small proportion of radiocarbon as a result. All plants and animals therefore contain a certain amount of radiocarbon along with the stable isotope ^{12}C. When a living thing dies, it stops taking in radiocarbon, and the radiocarbon it already contains decays steadily. By measuring the ratio between the ^{14}C and ^{12}C contents of the remains of an animal or plant and comparing it with the ratio of these isotopes in living organisms, the time that has passed since the death of the animal or plant can be found. (*a*) How old is a piece of wood from an ancient dwelling if its relative radiocarbon content is 1/4 that of a modern specimen? (*b*) If it is 1/16 that of a modern specimen?

(*a*) Since $1/4 = 1/2 \times 1/2$, the specimen is two half-lives old, which is 11,200 years old.

(*b*) Since $1/16 = 1/2 \times 1/2 \times 1/2 \times 1/2$, the specimen is four half-lives old, which is 22,400 years old.

12.13. **Why can radiocarbon dates never be completely reliable?**

The basic problem with radiocarbon dating is that the $^{14}C/^{12}C$ ratio in the atmosphere has not been constant throughout the past. Since about 1850, for instance, large amounts of carbon 12 have been added to the atmosphere by the burning of the fossil fuels coal and oil whose original carbon 14 content has long since decayed. More recently, nuclear explosions have doubled the carbon 14 concentration in the atmosphere. From a longer perspective, there is no reason to suppose that cosmic rays have always reached the earth at their present intensity. In fact, since the geomagnetic field deflects many cosmic rays back into space and this field is known to vary in strength, cosmic-ray intensity must also have varied in the past and with it the rate of formation of carbon 14. Comparisons between tree-ring counts and radiocarbon dates for trees several thousand years old show discrepancies of over 10%, with the radiocarbon dates being the younger.

12.14. **How can the temperatures of ancient seas and oceans be determined?**

Oxygen consists of several isotopes whose atoms have the same chemical behavior but are slightly different in mass. The relative abundances of the different isotopes in sea water depend upon the temperature of the water; the $^{18}O/^{16}O$ ratio, for instance, decreases with increasing temperature. Thus the temperature of an ancient body of water can be established by measuring the $^{18}O/^{16}O$ ratio in fossil shells of marine organisms that lived in that body of water.

12.15. **Precambrian rocks are exposed over a large part of eastern Canada. What does this observation suggest about the geological history of this region since the end of Precambrian time?**

The region must have been above sea level for most of the 570 million years since the end of Precambrian time or else it would be covered with sedimentary rocks.

12.16. **The early atmosphere of the earth probably consisted of carbon dioxide, water vapor, and nitrogen, with little free oxygen. What is believed to be the source of the oxygen in the present-day atmosphere? What bearing has this question on the relatively rapid development of varied and complex forms of life that marks the start of the Paleozoic era?**

Most of the oxygen in the atmosphere was probably produced by photosynthesis. In Precambrian time photosynthesis was carried out by the blue-green algae that were abundant then; these algae do not require free oxygen, unlike plants. When the oxygen content of the atmosphere and oceans grew large enough, more complex, oxygen-dependent forms of life could develop.

12.17. **Paleozoic sedimentary rocks derived from marine deposits are widely distributed in all the continents. What does this indicate about the height of the continents relative to sea level in the Paleozoic area?**

Much of the area of the continents must have been near or below sea level during at least part of the Paleozoic since shallow seas must have been widespread on their surfaces then.

12.18. **What are some of the chief differences between reptiles and mammals?**

Reptiles are cold-blooded (their body temperatures vary with the ambient temperature), lay eggs, have relatively small brains and scaly skins, and the teeth of each individual are all nearly the same. Mammals are warm-blooded (their body temperatures are constant), bear live offspring which are suckled, have relatively large brains and usually hairy coats and subcutaneous fat (both of which provide thermal insulation), and the teeth of each individual are of several different kinds.

12.19. **The same reptiles were found on all continents during the Mesozoic era, but the mammals of the Cenozoic era are often different on different continents. Why?**

During the Mesozoic era today's continents were joined together so the animal populations (which were largely reptiles) could move freely among them. During the Cenozoic era the continents were split apart, and the evolution of some of the mammals that replaced the reptiles proceeded differently on the various land masses.

12.20. How were the coral atolls of the Pacific Ocean formed?

Coral is composed of the remains of living organisms (called coral polyps) that live in shallow coastal waters in tropical regions. Figure 12-4a shows a volcanic island with a coral reef just offshore. An atoll forms if the island subsides or sea level rises; the reef is built upward by the coral polyps in order that they be near the water surface where the plankton on which they feed live, as in Fig. 12-4b. Eventually the original island may sink well below sea level to leave a narrow coral reef that fringes a central lagoon, as in Fig. 12-4c.

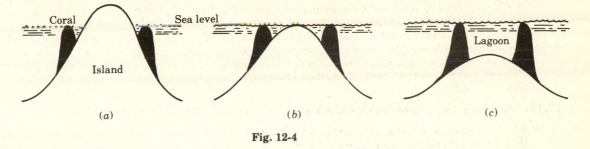

Fig. 12-4

12.21. Under what circumstances is coal formed? During what geological periods were such conditions widespread?

Coal is formed from plant material that accumulates in an environment where partial decay occurs in which most of the hydrogen and oxygen is removed to leave a residue that is largely carbon. The residue consolidates into coal under the pressure of sediments deposited later. Swamps are especially favorable for the formation of coal, since there is an abundance of plant life and the decay of plant remains underwater leaves carbon residues. Most coal deposits were laid down during the Mississippian and Pennsylvanian periods, often jointly called the Carboniferous period.

12.22. What is believed to be the origin of petroleum?

Petroleum is thought to have originated in the remains of marine animals and plants which accumulated on sea floors where the water circulation is so slow that little oxygen is present. Bacterial decomposition of the organic matter left residues that consisted chiefly of carbon and hydrogen. These residues were buried under sediments and under the influence of heat and pressure became the liquid and gaseous hydrocarbon compounds that constitute petroleum and natural gas respectively. The oil and gas migrated through porous beds until they either emerged at the surface or were trapped by an impermeable layer. Most petroleum and natural gas are found in Cenozoic and Mesozoic sandstones and carbonate rocks.

12.23. Give examples of geological structures that can act as traps for petroleum and natural gas.

About 80% of oil and natural gas deposits are found in anticline traps such as that shown in Fig. 12-5a. Since gas is lighter than oil and both are lighter than water, they are always found in this sequence from top to bottom. Fault and unconformity traps are also shown in Figs. 12-5b and c; still other types of traps are known.

12.24. What were the Ice Ages? When did they occur?

The Ice Ages involved the formation of ice sheets that covered large areas of the earth's surface. Ice advanced across the continents during four major episodes, which were separated by interglacial periods during which the ice retreated poleward. The Ice Ages took place during the past two million years, that is, in the Pleistocene epoch of the Quarternary period of the Cenozoic era. The most recent large-scale glaciation covered much of Canada and northeastern United States and began to recede only about 20,000 years ago. The origin of the worldwide climatic changes that produced the Ice Ages is not known, though a number of possible mechanisms have been proposed.

12.25. The Scandinavian land mass is rising at the rate of about 1 cm per year. What is believed to be the reason?

When the thick sheet of ice that covered this region during the most recent glacial period melted, the continental block became lighter and its buoyancy provided an upward force that has been raising it toward a level of isostatic equilibrium.

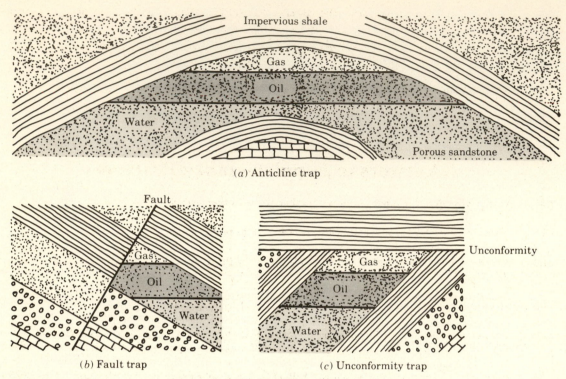

(a) Anticline trap

(b) Fault trap   (c) Unconformity trap

Fig. 12-5

Supplementary Problems

12.26.   What is an unconformity?

12.27.   Precambrian rocks include sedimentary, igneous, and metamorphic varieties.   What does this suggest about geological activity in Precambrian time?

12.28.   What conspicuous difference is there between Precambrian sedimentary rocks and those of later eras?

12.29.   What are the chief kinds of organisms that have left traces in Precambrian sedimentary rocks?

12.30.   Why are fossils never found in igneous rocks and only seldom in metamorphic rocks?

12.31.   Why is it believed that large parts of the United States were once covered by shallow seas?

12.32.   About 200 million years ago today's continents were all part of the supercontinent Pangaea.   During what geological era did Pangaea break apart into Laurasia and Gondwanaland?   During what era did Laurasia break up into North America, Greenland, and Eurasia?

12.33.   What major change in land animal life occurred between the late Mesozoic and early Cenozoic era?

12.34.   During what geological era did birds develop?   From what type of animal did they evolve?

12.35.   Minnesota has a great many shallow lakes.   How did they originate?

12.36.   What is the age of the oldest rocks known?   How was this age determined?

12.37.   The half-life of a radioactive isotope of hydrogen called *tritium* is 12.5 years.   If we start out with 1 g of tritium, how much will be left undecayed after 25 years?

12.38　　The half-life of a certain radioactive isotope of sodium is 15 hr.　How long does it take for $\frac{7}{8}$ of a sample of this isotope to decay?

12.39　　Why does the radiometric age of a metamorphic rock often refer to the time of its metamorphism rather than to the time the original rock was formed?

12.40.　　What procedure would you follow to find the age of an ancient piece of wood by radiocarbon dating?

Answers to Supplementary Problems

12.26.　An unconformity is an eroded surface buried under rocks that were deposited later.

12.27.　Precambrian geological activity must have been similar to that of today.

12.28.　Precambrian sedimentary rocks contain few if any fossils, whereas later sedimentary rocks usually contain abundant fossils.

12.29.　Bacteria and algae.

12.30.　Igneous rocks have hardened from a molten state, and no fossil could survive such temperatures.　Metamorphic rocks have been altered under conditions of heat and pressure severe enough to distort or destroy most fossils.

12.31.　Sedimentary rocks that contain the fossil shells of marine organisms are found in many parts of the United States.

12.32.　Mesocoic era; Cenozoic era.

12.33.　Reptiles, the dominant form of land animals during the Mesozoic, declined and were superseded by mammals, which are dominant in the Cenozoic.

12.34.　Mesozoic era; reptiles.

12.35.　The Pleistocene glaciation in that region left many depressions which subsequently filled with water to form lakes.

12.36.　Rocks 4 billion years old have been found in Greenland; their ages were established by radioactive dating.

12.37.　Since 25 years is two half-lives here, $\frac{1}{2} \times \frac{1}{2} = \frac{1}{4}$ of the original amount of tritium will be left, which is 0.25 g.

12.38.　After $\frac{7}{8}$ had decayed, $\frac{1}{8}$ is left, and $\frac{1}{8} = \frac{1}{2} \times \frac{1}{2} \times \frac{1}{2}$ which is 3 half-lives.　Hence the answer is 3×15 hr $= 45$ hr.

12.39.　In both uranium-lead decay series, one of the intermediate products is an isotope of the gas radon, and the stable daughter product of the decay of potassium 40 is an isotope of the gas argon..　During the metamorphism of a rock some of the original minerals recrystallize into other minerals, and the gas atoms present may not be incorporated in the new mineral grains.　The loss of the decay products from the mineral grains means in effect that the radioactive clock starts again from zero in such rocks at the time of metamorphism.

12.40.　The procedure is to determine the radioactivity of a known mass of carbon from the ancient wood and to compare it with the radioactivity of the same mass of carbon from a piece of wood of recent origin.　The difference between the two activities can be converted into an age figure for the ancient wood by taking into account the 5600-year half-life of radiocarbon.

Chapter 13

The Earth as a Globe

MOTIONS OF THE EARTH

The earth rotates on its axis and revolves around the sun. The *day* is the time required for a complete rotation and the *year* is the time required for a complete revolution. The axis of rotation is tilted by 23.5° with respect to a perpendicular to the plane of the orbit. The angle of tilt is constant but its direction in space changes very slowly; thus the earth resembles a wobbling top. The wobble is called *precession* and its period is about 26,000 years.

The *seasons* occur because, as a result of the tilt of the earth's axis, for half of each year one hemisphere receives more sunlight than the other, and in the other half of the year it receives less sunlight (Fig. 3-3). On 22 June (the *summer solstice*) the noon sun is at its highest in the sky in the northern hemisphere and the period of daylight is longest; on 22 December (the *winter solstice*) the noon sun is at its lowest and the period of daylight is shortest. In the southern hemisphere the situation is reversed. On 21 March (the *vernal equinox*) and 23 September (the *autumnal equinox*) the sun is directly overhead at noon on the equator and the periods of daylight and darkness are equal everywhere.

LATITUDE AND LONGITUDE

Locations on the earth's surface are specified in terms of its axis of rotation. A *great circle* is any circle on the earth's surface whose center is the earth's center. The *equator* is a great circle midway between the North and South Poles. A *meridian* is a great circle that passes through both poles, and it forms a right angle with the equator. The *prime meridian* passes through Greenwich, England. The *longitude* of a point on the earth's surface is the angular distance between a meridian through this point and the prime meridian; the prime meridian is assigned the longitude 0°, and longitudes are given in degrees east or west of the prime meridian. Thus a longitude of 60° W identifies a meridian 60° west of the prime meridian.

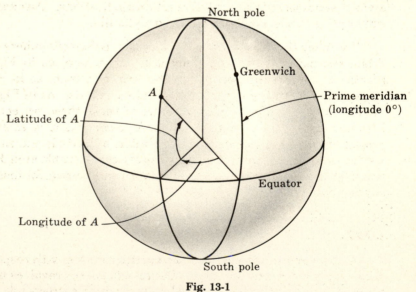

Fig. 13-1

The *latitude* of a point on the earth's surface is the angle between a line drawn from the earth's center to it and another line drawn from the earth's center to a point on the equator on the same meridian. Thus a latitude of 60° N identifies a circle (smaller than a great circle) 60° north of the equator. The latitude and longitude of a place on the earth's surface specify its location. (See Figs. 13-1 and 13-2.)

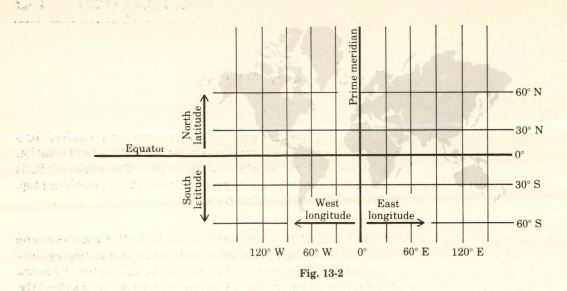

Fig. 13-2

MAPS

A *map* is a representation of part or all of the earth's surface on a sheet of paper. Because the earth is round and a sheet of paper is flat, some distortion must occur in preparing a map: the shapes of geographic features will not be correct, or the scale of areas will vary from place to place, or both.

A *map projection* is a particular method used to transfer parallels of latitude and meridians of longitude to a flat surface. On a *conformal* (or *orthomorphic*) projection, shapes are reproduced without distortion; on an *equal-area* projection, the scale of areas is the same everywhere. No map can be both conformal and equal-area, but for many purposes one or the other feature is all that is required. Some projections are neither conformal nor equal-area, but have other properties that make them especially suitable for certain applications. An example is the *gnomonic* projection on which great-circle arcs appear as straight lines.

There are three basic ways in which the surface of a sphere can be mapped on a sheet of paper. When the paper is flat and tangent to the sphere, as in Fig. 13-3a, the result is a *zenithal* projection. When the paper is in the form of a cone, as in Fig. 13-3b, the result is a *conical* projection. When the paper is rolled into a cylinder, as in Fig. 13-3c, the result is a *cylindrical* projection. All maps employ versions of these three projections. For example, the familiar *Mercator* projection of Fig. 13-2, whose development is shown in Fig. 13-3c, is a cylindrical projection in which the spacing of parallels of latitude increases from equator to poles so as to produce a conformal map at the expense of a considerable area distortion: on a Mercator map of the world Greenland appears as larger than South America, for instance, although it is actually many times smaller.

TIME

A *solar day* is the period of the earth's rotation with respect to the sun; a *sidereal day* is its period with respect to the stars. Because the earth revolves around the sun, the sidereal day is about 4 min shorter than the solar day, which corresponds to one day's difference per year. Ordinary timekeeping is based on the average solar day, which is divided into 24 hr with each hour further divided into minutes and seconds.

From the point of view of a stationary observer on the earth, the sun moves around the earth in a westward direction once very 24 hr, which is 15° of longitude per hour. This means that noon (or any other time reckoned with respect to the sun) occurs 1 hr later at a longitude 15° west of a particular place and 1 hr earlier at a longitude 15° east of it. For convenience, the world is divided

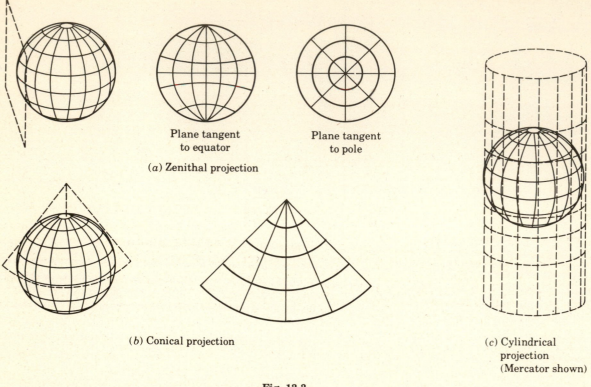

Plane tangent
to equator

Plane tangent
to pole

(a) Zenithal projection

(b) Conical projection

(c) Cylindrical
projection
(Mercator shown)

Fig. 13-3

into 24 *time zones,* each about 15° of longitude wide and each keeping time 1 hr ahead of the zone west of it and 1 hr behind the zone east of it.

The *international date line*, which follows the 180° meridian except for deviations to avoid going through Alaska and island groups in the Pacific, separates one day from the next. A person traveling eastward around the world sets his watch ahead at each successive time zone, and when he crosses the date line he subtracts a day from his calendar in order to compensate. A person traveling westward sets his watch behind at each successive time zone, and when he crosses the date line he adds a day to his calendar.

Solved Problems

13.1.   Why is the earth round?

Gravitational forces in the earth and other large astronomical bodies are strong enough to prevent more than minor departures from sphericity (except for the centrifugal distortion discussed in Problem 13.2). If a significant protuberance were to occur on the earth, the gravitational pull of the rest of the earth would lead to such overwhelming pressures on the underlying material in the earth's interior that it would flow out sideways until the protuberance became level or nearly level with the rest of the surface. Pressures under the margin of a large cavity would similarly cause the underlying material to flow into it. Such irregularities as mountains and ocean basins are on too small a scale to greatly distort the pressure balance in the interior.

13.2.   Apart from surface irregularities such as mountains, is the earth a perfect sphere?

Because the earth is spinning rapidly, its equatorial parts tend to swing outward, just as a ball on a string does when it is whirled around. As a result the earth bulges slightly at the equator and is slightly flattened at the poles, much like a grapefruit (Fig. 13-4); the difference between the polar and equatorial diameters is 43 km, which is 0.34%. The effect is known as *centrifugal distortion*.

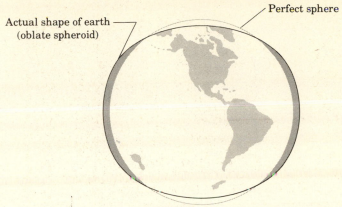

Actual shape of earth (oblate spheroid)

Perfect sphere

Fig. 13-4

13.3. **What is the difference between the nautical mile and the statute mile?**

The statute mile, equal to 5280 ft or 1.609 km, is a traditional unit of length in English-speaking countries. The nautical mile, equal to 6076 ft or 1.852 km, is very nearly the length of $1'$ of latitude anywhere on the earth. The nautical mile is especially convenient for air and sea navigation since distances in this unit can be read off on the latitude scale of a map.   A nautical mile equals 1.151 statute miles.   The *knot* is a velocity unit equal to 1 nautical mile/hr.

13.4. **How long is a degree of latitude?   Of longitude?**

The spacing of parallels of latitude is very nearly constant with $1° \simeq 60$ nautical miles $\simeq 69$ statute miles $\simeq 111$ km.   The deviations from constancy arise because the earth is not a perfect sphere; a degree of latitude is about 1% longer at the poles than at the equator.

The spacing of meridians of longitude varies from a maximum at the equator to 0 at the poles.   At the equator $1°$ of longitude is a trifle longer than $1°$ of latitude there; at the latitude of New York City (approximately $40°$ N), $1°$ of longitude = 85 km; and at the latitude of Anchorage, Alaska (approximately $60°$ N), $1°$ of longitude = 56 km.

13.5. **(a) How are the North and South Poles and the equator defined?   (b) What are their respective latitudes?**

(a)   The North and South Poles are the points where the earth's axis of rotation intersects its surface.   The equator is an imaginary line around the earth midway between the poles.

(b)   The latitude of the North Pole is $90°$ N, that of the South Pole is $90°$ S, and that of the equator is $0°$.

13.6. **(a) How are the Arctic and Antarctic Circles defined?   (b) What are their latitudes?**

(a)   On 22 December, the shortest day of the year in the northern hemisphere, the $23.5°$ tilt of the earth's axis means that no sunlight reaches any point within $23.5°$ of the North Pole.   The Arctic Circle is the boundary of this region of darkness.   On the same day, which is the longest day of the year in the southern hemisphere, there are 24 hours of daylight at all points within $23.5°$ of the South Pole, and the Antarctic Circle is the boundary of this region of daylight.   On 22 June the situations in the two hemispheres are reversed.

(b)   The latitude of the North Pole is $90°$ N, hence that of the Arctic Circle is $90°$ N $- 23.5° = 66.5°$ N.   Similarly the latitude of the Antarctic Circle is $66.5°$ S (Fig. 13-5).

13.7. **(a) How are the Tropics of Cancer and Capricorn defined?   (b) What are their latitudes?**

(a)   The Tropic of Cancer is the most northerly latitude in the northern hemisphere at which the sun is ever directly overhead at noon.   The Tropic of Capricorn is the corresponding latitude in the southern hemisphere.

(b)   On 22 June, when the North Pole is tilted closest to the sun and hence is the day of maximum sunlight in the northern hemisphere, the noon sun is directly overhead $23.5°$ north of the equator; hence the latitude of the Tropic of Cancer is $23.5°$ N. Similarly the latitude of the Tropic of Capricorn is $23.5°$ S; the South Pole is tilted closest to the sun on 22 December, when the noon sun is directly overhead at this latitude (Fig. 13-5).

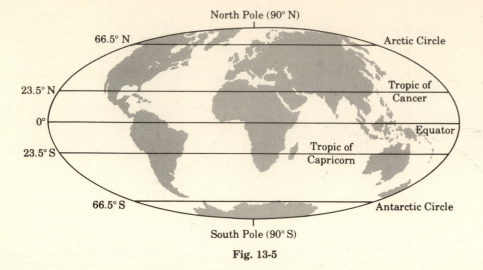

North Pole (90° N)

66.5° N   Arctic Circle

23.5° N   Tropic of
                                                                              Cancer

0°   Equator

23.5° S   Tropic of
                                                                              Capricorn

66.5° S   Antarctic Circle

South Pole (90° S)

Fig. 13-5

13.8. Where on the earth can the entire sky be observed during the course of a year?

    The entire sky can be seen, though not all of it at the same time, between the Tropics of Cancer and Capricorn, which are the limiting latitudes that intersect the plane of the earth's orbit.

13.9. As seen from the earth, the sun drifts eastward relative to the stars; that is, at sunset or sunrise on a given day, the sun appears eastward of its position the previous sunset or sunrise. Through approximately what angle does the sun move eastward each day relative to the stars?

    Since the earth takes a year to revolve counterclockwise (as seen from above the North Pole) around the sun, the sun appears to an observer on the earth to drift eastward relative to the stars through 360° in 365 days. The daily drift is therefore 360°/365 days, which is a little less than 1°/day.

13.10. Polaris, the North Star, is almost directly over the North Pole today. Was it always there? Will it remain there in the future?

    The direction in space of the earth's axis gradually changes, an effect called *precession*. Hence in the past Polaris was not as close to being directly over the North Pole as it is today and in the future it will diverge farther and farther from being overhead there. The period of the earth's precession is about 26,000 years, and in 26,000 years from now Polaris will again be the North Star.

13.11. What is the *declination* of the sun?

    The declination of the sun is the latitude at which it is vertically overhead at a particular time. The sun's declination varies from 23.5° N at the summer solstice to 23.5° S at the winter solstice; it is 0° at the equinoxes when the sun is overhead at the equator.

13.12. A sextant is a device for finding the angular altitude of a celestial body above the horizon. How can a sextant measurement of the sun's altitude at local noon, when it is at the highest point in the sky, be used to find the latitude of the place of observation?

    The sextant altitude h must first be converted into the *zenith angle* θ between the sun's position in the sky and the zenith, which is the point directly overhead the observer. As in Fig. 13-6a, $\theta = 90° - h$. From the diagram in Fig. 13-6b it is clear that, if the sun is south of the observer's zenith and the declination is north, or vice versa, the latitude equals θ plus the sun's declination (which can be obtained from tables in an almanac). If the sun is north of the observer's zenith and the declination is north, or if both are south, the latitude equals the difference between θ and the declination.

13.13. What do contour lines on a map represent?

    Each contour line on a map connects points whose elevation above sea level is the same. Figure 13-7 shows how a hill is represented by contour lines. It is customary to draw contour lines in even steps of vertical distance on a particular map. The more closely spaced contour lines are in a certain region on a map, the steeper the slope is there.

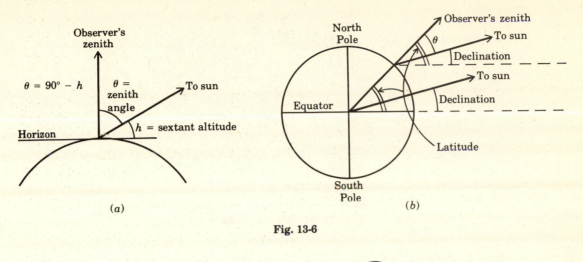

(a)   (b)

Fig. 13-6

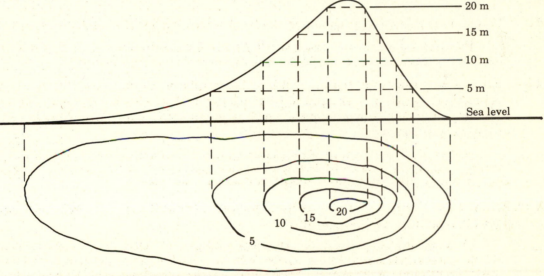

Fig. 13-7

13.14. A *chart* is a map intended for navigation. Nearly all charts use the Mercator projection even though other conformal projections exist whose scale varies less with latitude. Why?

The Mercator projection is the only one on which a straight line represents a path of constant compass direction (if the compass is a magnetic one rather than a gyrocompass, its reading must be corrected for local magnetic variation). Such a path is called a *rhumb line*, or *loxodrome*. To determine the course to follow from one place to another, a navigator need only draw a straight line joining the two on a Mercator chart and use a protractor to find the direction of the resulting rhumb line relative to either true north or magnetic north.

13.15. The shortest distance between two places on the earth's surface is an arc of a great circle. How can a navigator determine a great-circle path? How does such a path appear on a Mercator chart?

The most direct way to find the great-circle path between two points is to use a globe. In practice, navigators use a gnomonic map instead. If a light bulb is placed at the center of a transparent globe, the projection of the globe's surface on a plane held tangent to the globe will be a gnomonic map. All straight lines on such a map are great circles. The great-circle path between two places on the earth's surface can be drawn on a Mercator chart by transferring a number of intermediate points from a gnomonic map (or directly from a globe) and then connecting them. The result will be a series of line segments that approximates the curved line that a great-circle path should appear as on a Mercator chart. Although a great-circle path seems longer on a Mercator chart than a rhumb line, it is actually shorter in distance over the earth's surface.

13.16. The Lambert conformal conic projection is very widely used. What is the basis of this projection?

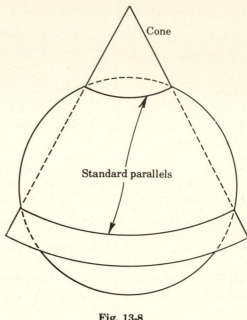

Cone

Standard parallels

On a Lambert map, the cone on which the projection is made intersects the globe at two parallels of latitude, as in Fig. 13-8, instead of at just one parallel as in Fig. 13-3b. The advantage of a Lambert map is that, when the spacing of the various parallels of latitude is adjusted to make the map conformal, the scale varies less with latitude than it does on a tangent-cone map. On a Lambert map, as on other conic maps, the meridians are straight lines that radiate from a common point and the parallels are circular arcs.

13.17. (a) Why does the sky not become dark as soon as the sun sets below the horizon? (b) Is the duration of twilight the same everywhere on the earth?

(a) The scattering of sunlight by air molecules, moisture droplets, and dust particles in the earth's atmosphere enables some sunlight to be reflected back to the earth's surface for an hour or more after the sun has set.

(b) Twilight is briefest at the equator where the sun rises and sets perpendicular to the horizon. At higher latitudes it rises and sets at shallower angles, thereby extending the period of twilight.

Fig. 13-8

13.18. Why does the duration of the solar day vary slightly throughout the year?

The earth's orbit is an ellipse, and its velocity is greater when it is near the sun than when it is far from the sun. Since the solar day depends on both the earth's rotation and its orbital motion, the duration of the solar day varies during the year.

13.19. *Greenwich mean time* (GMT) is local time at the prime meridian (0° longitude) when variations in the earth's orbital velocity are averaged out. When GMT is 0800 (8:00 AM), what is the local time at (a) New York City (longitude 74° W) and (b) Venice (longitude 12° E)?

(a) Since a longitude difference of 15 leads to a time difference of 1 hr, 1° of longitude difference is equivalent to a 4 min time difference. The longitude of New York City is 74° W, hence the time difference with respect to GMT is

$$\Delta T = 74° \times 4 \text{ min/°} = 296 \text{ min} = 4 \text{ hr } 56 \text{ min}$$

Since New York City is west of the prime meridian, the time difference must be subtracted from GMT (it is earlier in New York City than in Greenwich), so local time in New York City at 0800 GMT is

$$T = \text{GMT} - \Delta T = 0800 - 0456 = 0760 - 0456 = 0304$$

which is 3:04 AM.

(b) The longitude of Venice is 12° E, hence the time difference is

$$\Delta T = 12° \times 4 \text{ min/°} = 48 \text{ min}$$

Since Venice is east of the prime meridian, the time difference must be added to GMT (it is later in Venice than in Greenwich), so local time in Venice at 0800 GMT is

$$T = \text{GMT} + \Delta T = 0800 + 0048 = 0848$$

which is 8:48 AM.

13.20. What is the GMT of local noon at Lisbon, longitude 9° W?

Since 1° of longitude is equivalent to 4 min of time, the time difference between 0° and 9° longitude is

$$\Delta T = 9° \times 4 \text{ min}/° = 36 \text{ min}$$

Lisbon is west of the prime meridian, so when it is $T = 1200$ at Lisbon the GMT is

$$\text{GMT} = T + \Delta T = 1200 + 0036 = 1236$$

which is 12:36 PM.

13.21. When the local time in Moscow (longitude 38° E) is 1000 (10:00 AM), what is the local time in New York City (longitude 74° W)?

The longitude difference between Moscow and New York City is 38° + 74° = 112° since they are on opposite sides of the prime meridian. The corresponding time difference is

$$\Delta T - 112° \times 4 \text{ min}/° = 448 \text{ min} = 7 \text{ hr } 28 \text{ min}$$

Since New York City is west of Moscow, local time in New York City is earlier than local time in Moscow, and the New York City local time that corresponds to 1000 Moscow time is

$$T = 1000 - 0728 = 0960 - 0728 = 0232$$

which is 2:32 AM.

13.22. What is your longitude if local noon occurs at 1440 GMT (2:40 PM)?

The time difference is 2 hr 40 min = 160 min, and the corresponding longitude difference is

$$\frac{160 \text{ min}}{4 \text{ min}/°} = 40°$$

Since local noon occurs *after* 1200 GMT, your location must be west of the prime meridian, so the longitude is 40° W.

13.23. Why are leap years necessary?

The length of the year is 365 days 5 hr 48 min 46 s. Since the difference between the actual length of the year and 365 days is very nearly 6 hr, which is $\frac{1}{4}$ day, adding an extra day to February every four years (namely those years evenly divisible by 4) enables the seasons to recur at very nearly the same dates each year. (The discrepancy of 11 min 14 s per year that remains adds up to a full day after 128 years. To remove most of this discrepancy, century years not divisible by 400 are not leap years; thus 1900 was not a leap year, but 2000 will be one. This step makes the calendar accurate to 1 day per 3300 years. A further modification leaves 4000, 8000, 12,000 and so on as 365-day years rather than leap years; the resulting calendar is accurate to 1 day per 20,000 years, which is adequate for the time being.)

13.24. The dates of the equinoxes and solstices are not always the same from year to year; the vernal equinox, for instance, may occur on either March 20 or 21. Why should there be any variation in these dates?

The length of the year is very nearly 365¼ days, and the difference between the true year and the calendar year therefore increases to a full day every four years when the addition of an extra day to February to make a leap year corrects the discrepancy. For this reason the equinoxes and solstices, which follow the true 365¼-day year, may occur on different dates on various calendar years.

Supplementary Problems

13.25. (a) If the earth's axis were tilted by 30° instead of by 23.5°, would the seasons be more or less pronounced than they now are? (b) What would the latitudes of the Arctic Circle and the Tropic of Cancer be?

13.26.   (a) In May, does the length of the day (that is, the period between sunrise and sunset) change when one travels north from the Tropic of Cancer? If so, does it become longer or shorter? (b) Does the length of the day change when one travels west from the prime meridian? If so, does it become longer or shorter?

13.27.   For how long each year does the sun remain below the horizon at the Arctic Circle? At the North Pole?

13.28.   How does the path of the sun in the sky appear to an observer at the equator at an equinox?

13.29.   How does the path of the sun in the sky appear to an observer at the North Pole during the year?

13.30.   The simplest map projection is a rectangular grid with equally-spaced latitude and longitude lines. What class of projection is this? What are its advantages and disadvantages?

13.31.   In a transverse Mercator projection, the cylinder of paper upon which the globe is imagined to be projected has its axis perpendicular to the earth's axis instead of being coincident with it as in the usual Mercator projecton. What is the advantage of this projection?

13.32.   How does a stream valley appear on a contour map?

13.33.   What is the local time in Tokyo (longitude 140° E) when it is local noon in Moscow (longitude 38° E)?

13.34.   When GMT is 1430 (2:30 PM), what is the local time in Chicago, longitude 88° W?

13.35.   What is GMT when it is 1800 (6:00 PM) local time in Tokyo, longitude 140° E?

13.36.   What is your longitude if local noon occurs at 0800 GMT (8:00 AM)?

13.37.   Can you think of any evidence in favor of the earth's rotation that does not involve any reference to astronomical bodies outside the earth?

13.38.   Jupiter is much larger than the earth but its period of rotation is less than 10 hr as compared with 24 hr for the earth. How would you expect Jupiter's centrifugal distortion to compare with that of the earth?

Answers to Supplementary Problems

13.25.   More pronounced; 60°, 30°.

13.26.   The day becomes longer; no change.

13.27.   One day; six months.

13.28.   The sun appears to move in a vertical semicircle from the eastern horizon at dawn through a point directly overhead at noon to the western horizon at sunset.

13.29.   The path of the sun appears as a spiral making one turn per day that rises above the horizon at the vernal equinox, continues climbing until the summer solstice, and then descends in a spiral until it disappears below the horizon at the autumnal equinox.

13.30.   Such a projection is cylindrical. A map using this projection is very easy to prepare, and is adequate when a region a few miles across is involved. The projection is neither conformal nor equal-area, however, and leads to severe distortions when large regions are to be mapped.

13.31.   Both polar regions can be shown on a transverse Mercator map; this is impossible on the usual Mercator projection where the spacing of parallels of latitude approaches infinity as the poles are approached.

13.32.   Each contour line that crosses a stream valley has a V-shaped kink that points upstream.   The apex of the V lies on the stream bed.   The reason for these kinks is that each contour follows a certain elevation along the side of the valley upstream to the point where the stream has that elevation.

13.33.   1848 (6:48 PM).

13.34.   0838 (8:38 AM).

13.35.   0840 (8:40 AM).

13.36.   60° E.

13.37.   The behavior of a Foucault pendulum; the equatorial bulge; the pattern of winds in the general circulation of the atmosphere.

13.38.   It is more pronounced.

Chapter 14

The Solar System

PTOLEMAIC AND COPERNICAN SYSTEMS

Until the 16th century the earth was believed to be the center of the universe. According to the *Ptolemaic system,* which was a detailed picture of this idea, the sun and moon revolve around the earth in circular orbits while the planets each travel in a series of loops called *epicycles.* The stars are supposed to be fixed to a crystal sphere that turns once a day.

In 1543 Copernicus published the hypothesis that the sun is the center of the solar system, with the moon revolving around the earth. The stars are far away in space, and the earth rotates daily on its axis. In the *Copernican system* the planet nearest the sun is Mercury and next in order are Venus, the earth, Mars, Jupiter, Saturn, Uranus, Neptune, and Pluto. (The last three planets were not known in the time of Copernicus.)

Kepler modified the Copernican system by showing that the planetary orbits are ellipses rather than circles and discovered two other significant regularities obeyed by the planets. His three laws are as follows:

1. The orbits of the planets around the sun are ellipses.

2. Each planet moves so that a line drawn from the planet to the sun sweeps out equal areas in equal times. Thus the planet moves most rapidly when it is closest to the sun and least rapidly when it is farthest from the sun.

3. The ratio between the square of the time required by a planet to revolve around the sun and the cube of its average distance from the sun has the same value for all the planets.

Figure 14-1 shows the arrangement of the planetary orbits. The actual orbits are ellipses rather than circles and their spacing in the diagram is not to scale, but the orbits do lie very nearly in the same plane.

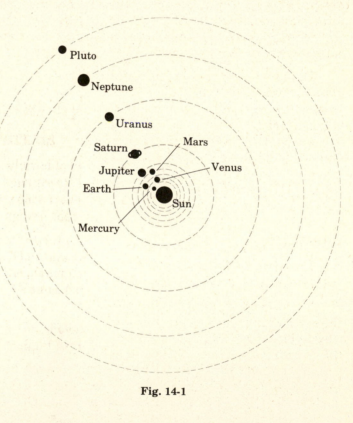

Fig. 14-1

GRAVITATION

The sun must be exerting attractive forces on the planets in order to hold them in their orbits. Isaac Newton proposed that these forces are examples of a quite general interaction called *gravitation* that occurs between all bodies in the universe.   Newton was able to infer his *law of universal gravitation* from Kepler's laws: Every body in the universe attracts every other body with a force that is directly proportional to each of their masses and inversely proportional to the square of the distance between them.   In equation form,

$$\text{Gravitational force} = G\frac{m_1 m_2}{r^2}$$

where m_1 and m_2 are the masses of any two bodies, r is the distance between them, and G is a constant whose value is the same everywhere in the universe.   A spherical body behaves gravitationally as though its entire mass were concentrated at its center.

Gravitational forces vary considerably with distance because of the $1/r^2$ factor.   Thus the gravitational force on a planet would drop to $\frac{1}{4}$ its original amount if the distance of the planet from the sun were to be doubled; if the distance were to be halved, the force would increase to 4 times its original amount.

THE PLANETS

Like the earth, the other planets rotate on their axes and revolve around the sun, and all except Mercury, Venus, and Pluto have satellites.   Most planet and satellite orbits lie near the same plane, and most of the various rotations and revolutions are in the same direction (counterclockwise as seen looking down from above the earth's North Pole).   The planets and their satellites are visible by virtue of the sunlight they reflect.

The inner planets—Mercury, Venus, the earth, and Mars—are considerably smaller, less massive, and denser than the outer planets—Jupiter, Saturn, Uranus, and Neptune; also, the inner planets rotate much more slowly than do the outer planets.   Pluto is a special case, with properties closer to those of the inner planets despite its distance from the sun.   The outer planets (except Pluto) apparently consist largely of hydrogen and hydrogen compounds such as methane and ammonia, which accounts for their low densities.

Thousands of minor planets called *asteroids* orbit the sun between Mars and Jupiter.   The largest asteroid is not quite 500 mi in diameter; most are much smaller.

THE MOON

The moon revolves around the earth once every 27.3 days.   Because the moon rotates on its axis with the same period, the same lunar hemisphere always faces the earth.

As the moon revolves around the earth, the extent of its illuminated hemisphere visible from the earth changes, which accounts for the *phases* of the moon.   When the moon is on the opposite side of the earth from the sun, sunlight reaches the entire lunar hemisphere facing the earth, and the resulting bright disk is called *full moon*.   At *new moon* the moon is between the earth and the sun, and the hemisphere facing the earth receives no sunlight and appears dark.   At intermediate positions in the lunar orbit different portions of the illuminated part of the moon's surface are visible (Fig. 14-2).   Owing to the motion of the earth around the sun, the period between full moons is 29.5 days, which is longer than the 27.3-day orbital period of the moon.

The plane of the moon's orbit is slightly tilted relative to the plane of the earth's orbit, which is why sunlight normally is able to reach the moon when it is on the opposite side of the earth from the sun, and why the moon normally does not obstruct the sun at new moon.   On certain occasions, however, earth, sun, and moon lie along a straight line, and *eclipses* then take place (Fig. 14-3).   During a *lunar eclipse* the earth is between the sun and the moon and its shadow falls on the moon; during a *solar eclipse* the moon is between the sun and the earth and its shadow falls on the earth.   Total eclipses occur because, although their actual diameters and their distances from the

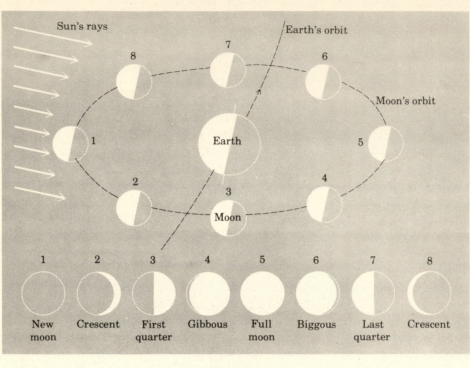

Fig. 14-2

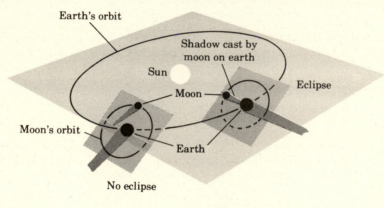

Fig. 14-3

earth are different, the apparent diameters of both sun and moon as seen from the earth are the same at certain times.

COMETS

Comets are members of the solar system whose orbits around the sun are very long, narrow ellipses. Most comet orbits are so large that their periods range up to a million years or more; a few have periods short enough to be seen regularly from the earth, for instance Halley's comet whose period is 76 years.

When it is far from the sun, a comet is a compact aggregate of frozen water, ammonia, and methane with some particles of metallic and stony characters probably present as well. Near the sun the H_2O, NH_3, and CH_4 vaporize and spread out to form a cloud of thin gas tens or hundreds of thousands of miles across. This cloud is excited by solar ultraviolet radiation and reradiates visible light; a small part of the glow of a comet is reflected sunlight.   Such clouds often exhibit tails that always point away from the sun; they are the result of pressure exerted both by sunlight and by the *solar wind* of protons and electrons that continually stream outward from the sun.

METEORS

Meteoroids are particles of matter, usually small, that travel through the solar system in orbits around the sun.   When a meteoroid enters the atmosphere, it is heated by friction and glows brightly: the visual phenomenon is called a *meteor*.   *Meteorites* are the remains of meteoroids that reach the ground.   Most meteoroids occur in swarms and produce meteor showers at regular intervals.   Such meteoroids are thought to be the debris of comets, and in some cases their orbits can be identified as those of former comets.   Meteorites are divided into two main classes: *iron meteorites*, which are largely iron with some nickel also present, and *stony meteorites*, which consist chiefly of silicate minerals in a characteristic structure.   A few meteorites are composed of mixtures of iron and stony material.

Solved Problems

14.1.   Did Copernicus prove that his system is correct and the Ptolemaic system is incorrect?   If not, why is the Copernican system accepted today?

The Copernican system, by referring the planetary motions to the sun rather than to the earth, was much simpler than the Ptolemaic system and, when modified by Kepler, was more accurate in describing these motions. However, the Ptolemaic system could also be modified to achieve as much accuracy as the Copernican system, though in a much more complicated way.   The Copernican system, when it was proposed, was better than the Ptolemaic system because of its simplicity, but this simplicity did not make the Ptolemaic system wrong since all it involved was a shift in the choice of the reference point for reckoning planetary motions.   The Copernican system is today considered correct (and the Ptolemaic system incorrect) because there is direct experimental evidence for the motions of the planets around the sun; for example, the change in apparent position of nearby stars relative to distant stars as the earth revolves around the sun.

14.2.   Why are Mercury and Venus always seen either around sunset or around sunrise?

Mercury and Venus are closer to the sun than the earth is, hence an observer on the earth always sees them in the vicinity of the sun.   When one of them is east of the sun, it disappears below the horizon after the sun and is visible in the early evening; when it is west of the sun, it rises above the horizon before the sun and is visible in the early morning.

14.3.   The *astronomical unit* (AU) is the average radius of the earth's orbit; it is equal to 1.495×10^{11} m, which is about 93,000,000 miles.   (*a*) Find the value of the constant ratio in Kepler's third law in terms of AU and years.   (*b*) The average radius of Jupiter's orbit is 5.2 AU.   How long does Jupiter require to complete a revolution around the sun?

(*a*) If the average radius of a planet's orbit is R and its period of revolution is T, Kepler's third law states that T^2/R^3 has the same value for all the planets.   For the earth, $T = 1$ yr and $R = 1$ AU, so $T^2/R^3 = C = 1$ yr^2/AU^3.

(*b*) For Jupiter, $R = 5.2$ AU, and so

$$T^2 = CR^3 = 1\,\frac{\text{yr}^2}{\text{AU}^3} \times (5.2 \text{ AU})^3 = 141 \text{ yr}^2$$

$$T = \sqrt{141 \text{ yr}^2} = 11.9 \text{ yr}$$

14.4.   Neptune's period of revolution is 165 yr.   (*a*) What is the average radius of Neptune's orbit in AU?   (*b*) In miles?

(a) According to Kepler's third law, $T^2/R^3 = C$, hence $R^3 = T^2/C$. With $T = 165$ yr and $C = 1$ yr^2/AU^3 we have

$$R^3 = \frac{T^2}{C} = \frac{(165 \text{ yr})^2}{1 \text{ yr}^2/\text{AU}^3} = 2.72 \times 10^4 \text{ AU}^3$$

$$R = \sqrt[3]{2.72 \times 10^4 \text{ AU}^3}$$

To find this cube root, we begin by rewriting 2.72×10^4 as 27.2×10^3 since $\sqrt[3]{10^3} = 10$. Because $\sqrt[3]{\text{AU}^3} =$ AU as well, we obtain

$$R = \sqrt[3]{2.72 \times 10^4 \text{ AU}^3} = \sqrt[3]{27.2 \times 10^3} \text{ AU} = \sqrt[3]{27.2} \times \sqrt[3]{10^3} \text{ AU} = \sqrt[3]{27.2} \times 10 \text{ AU}$$

For an approximate result, we note that $3 \times 3 \times 3 = 27$, so $\sqrt[3]{27.2} \approx 3$. Numerical tables give the more exact result

$$R = 3.01 \times 10 \text{ AU} = 30.1 \text{ AU}$$

(b) Since 1 AU $= 9.3 \times 10^7$ mi, $R = 30.1$ AU $\times 9.3 \times 10^7$ mi/AU $= 2.8 \times 10^9$ mi.

14.5. Only the gravitational interaction influences the motions of the planets around the sun. Why are the other fundamental interactions (see Problem 1.4) not involved as well?

The strong and weak nuclear interactions are too limited in range to affect planetary motion. As for the electromagnetic interaction, because like charges repel and unlike charges attract, it is very difficult to separate neutral matter into large-scale assemblies of opposite charge, hence all astronomical bodies are electrically neutral.

14.6. The moon's mass is approximately 1% of the earth's mass. How does the gravitational pull of the earth on the moon compare with the gravitational pull of the moon on the earth?

The forces between two interacting bodies are always equal in magnitude and opposite in direction; this physical principle is known as Newton's third law of motion. Hence the gravitational pull of the earth on the moon is equal and opposite to the pull of the moon on the earth. The equality of the two forces is incorporated in the formula $F = GM_1 m_2/r^2$ for gravitational force, where it is obvious that it makes no difference which body is called 1 and which body 2.

14.7. The gravitational force of the earth on a dropped stone causes it to fall to the ground, but the moon does not fall to the ground despite the gravitational force of the earth on it. Why not?

The moon's orbital speed is sufficiently great to keep it moving in an elliptical path in which its tendency to "fall" toward the earth (a consequence of gravitation) is exactly balanced by its tendency to continue in motion along a straight line (a consequence of its momentum). If a stone were thrown horizontally with sufficient speed, it would become a satellite of the earth like the moon instead of falling to the ground.

14.8. The *weight* of a body is the gravitational force with which the earth attracts it. If a person weighs 160 lb, this means that the earth pulls him down with a force of 160 lb. Weight is different from mass, which is a measure of the quantity of matter in a body as determined by its response to an applied force. The weight of a body varies with its location near the earth (or other astronomical body) whereas its mass is the same everywhere in the universe. What would a person who weighs 160 lb at the earth's surface weigh at a height above the surface of one earth radius (6400 km)?

A height of 1 earth radius above the earth's surface means a distance of 2 earth radii from its center. Since gravitational forces vary as $1/r^2$, the gravitational force on the person is reduced to $1/(2)^2 = \frac{1}{4}$ of its value at the surface, which is $(160 \text{ lb})/4 = 40$ lb.

14.9. A hole is drilled to the center of the earth and a stone is dropped into it. When the stone is at the earth's center, how do its mass and weight compare with their values at the surface?

The stone's mass is an intrinsic property that is the same everywhere.   When the stone is at the earth's center, the gravitational forces of the matter around it cancel one another out, and its weight is therefore zero there.

14.10. Stars, planets, and satellites are all very nearly spherical in form, but some of the smaller asteroids have irregular shapes.   Explain these observations.

As discussed in Problem 13.1, gravitational forces are responsible for the sphericity of large astronomical bodies.   In the case of a small asteroid, which might be only a few miles across, the gravitational forces are much smaller and the rigidity of their material may well be sufficient for an irregular shape to be maintained.

14.11. (a) How is it possible to distinguish the planets from the stars by observations with the naked eye?   (b) By observations with a telescope?

(a)   When viewed over a period of time, a planet will be seen to change its position in the sky relative to the stars.   (This is the reason for the name "planet," which is Greek for "wanderer.")

(b)   Seen through a sufficiently powerful telescope, the planets appear as disks whereas the stars, which are much more distant, appear as points of light.

14.12. Why is Venus a brighter object in the sky than Mars?

Venus is closer to the sun than Mars and hence receives more sunlight to reflect.   It is larger than Mars, so the reflecting surface is greater in area.   Venus is surrounded by clouds whereas Mars has none, and these clouds constitute a better reflector of sunlight than the Martian surface; the white polar caps on Mars are too small to make much difference in this respect.   As a result of all these factors, Venus is not only brighter than Mars but is also at times the brightest object in the sky after the sun and moon.

14.13. Which planets would you expect to show phases like those of the moon?

Mercury and Venus, because their orbits are closer to the sun than that of the earth.

14.14. What is the nature of Saturn's rings?

The rings of Saturn consist of large numbers of small particles which orbit the planet like miniature satellites.

14.15. Would you expect the rings of Saturn to revolve with the same period, like parts of the same phonograph record?

No.   Each particle in the rings pursues its own orbit around Saturn, and by Kepler's third law the inner particles must have shorter periods than the outer particles.

14.16. (a) Why is the average surface temperature of Mars lower than that of the earth?   (b) Why do Martian temperatures vary between day and night to a greater extent than they do on the earth?

(a)   Mars is about 50% farther from the sun than the earth is and hence receives less solar radiation.

(b)   The Martian atmosphere is much thinner than that of the earth and hence is less able to absorb and retain energy radiated from the surface during the night.

14.17. As seen from the earth, the moon drifts eastward relative to the stars; that is, on a given night the moon appears eastward of its position the night before at the same time. Through what angle does the moon move eastward each day relative to the stars?

The moon circles the earth in 27.3 days relative to the stars, hence it travels through 360° in 27.3 days or

$$\frac{360°}{27.3 \text{ days}} = 13°/\text{day}$$

14.18. How long a time elapses between the moon at first quarter (when it appears as a half moon) and the full moon?

From first quarter to full moon is $\frac{1}{4}$ of a lunar cycle. Since the complete cycle takes 29.5 days, here we have $\frac{1}{4} \times 29.5$ days = 7.4 days.

14.19. Why is it believed that the moon's interior is different in composition from the earth's interior?

The average density of the moon is 3.3 g/cm³ whereas that of the earth is 5.5 g/cm³. Part of the reason for the smaller density of the moon is its smaller total mass, which means that pressures in its interior are less than those in the earth's interior. However, this factor is not sufficient to account for the large difference in densities. Possible reasons include a smaller proportion of iron in the moon's interior than in the earth's, and the presence of large quantities of low-density substances such as graphite.

14.20. The densities of rocks on the moon's surface are about the same as the density of the moon as a whole. What does this observation suggest about the thermal history of the moon?

Because the moon apparently has a more or less uniform density, it probably never was entirely molten as the earth was. If the moon had melted all the way through, its heavier constituents would have become concentrated by gravity in a core near the center with the lighter constituents forming a mantle around it, as in the case of the earth.

14.21. List some theories of the moon's origin. Are any of them free from serious objections?

(1) The moon was initially part of the earth and split off from it to become an independent body. (2) The moon was formed elsewhere in the solar system and later was captured by the earth's gravitational field. (3) The moon and the earth came into being together as a double-planet system.

There are some serious objections to each of these theories.

14.22. When a comet is close enough to the sun to be seen from the earth, stars are visible through both the comet's head and tail. What does this imply about the danger to the earth from a collision with a comet?

The density of a comet is extremely low when it is in the vicinity of the earth, and in a collision the comet material would simply be absorbed in the upper atmosphere.

14.23. Some meteor showers recur each year at about the same time, for instance the Perseid shower that appears early every August. Does this mean that the orbits of the meteoroids in the Perseid swarm all have periods of exactly one year?

No. The annual occurrence of a meteor shower at a particular date simply means that the earth's orbit intersects the common orbit of a meteoroid swarm at the particular point corresponding to that date. (The Perseid meteoroids are believed to be the remnants of a comet whose period was 105 years.) If the number of meteors seen is about the same each year, the meteoroids must be spread out along their common orbit; if the number varies considerably, the meteoroids must be bunched together.

14.24. Over ninety percent of the meteorites found after a known fall are stony, yet most of the meteorites in museums are iron. Why?

Stony meteorites resemble ordinary rocks whereas iron ones are conspicuously different; also, stony meteorites are more readily weathered than iron meteorites.

Supplementary Problems

14.25. A person takes a sensitive scale to a mountain top. If his mass does not change, will his weight appear to be more, less, or the same as it did at sea level?

14.26. A girl weighs 128 lb on the earth's surface. What would she weigh at a height above the surface of 3 earth radii?

14.27. The average radius of Mercury's orbit is 0.387 AU. (*a*) How many meters is this? (*b*) Use Kepler's third law to find Mercury's period of revolution around the sun.

14.28. The period of revolution of Uranus is 84 yr. Use Kepler's third law to find the average radius of its orbit in AU.

14.29. What are the largest and smallest planets? The planets nearest to and farthest from the sun?

14.30. Which of the planets are readily visible with the naked eye?

14.31. On which planet is the length of the year shortest? On which planet is it longest?

14.32. According to *Bode's law*, which was discovered two centuries ago by Titius, the mean orbital radii of the planets in AU are given by the formula $R_n = 0.4 + 0.3 \times 2^n$, where $n = 0$ for Venus, $n = 1$ for the earth, and so on; $R = 0.4$ for Mercury. Compare the predictions of Bode's law with the actual orbital radii in AU. Include the asteroids, whose mean orbital radii average 2.9 AU.

14.33. Approximately how many days elapse between new moon and full moon?

14.34. (*a*) In what phase must the moon be at the time of a solar eclipse? (*b*) At the time of a lunar eclipse?

14.35. If the moon were smaller than it is, would total eclipses of the sun still occur? Would total eclipses of the moon still occur?

14.36. If the earth had no atmosphere, would comets still be visible from its surface? Would meteoroids?

14.37. Why do comets have tails only in the vicinity of the sun? Why do these tails always point away from the sun, even when the comet is receding from it?

14.38. Is the likelihood of being struck by a meteoroid greater or less on the surface of Mars than on the earth's surface?

Answers to Supplementary Problems

14.25. His weight will be less than at the earth's surface because he is farther from the center of the earth.

14.26. $128 \text{ lb}/(4)^2 = 128 \text{ lb}/16 = 8$ lb.

14.27. 5.79×10^{10} m; 0.241 years.

14.28. 19 AU

14.29. Jupiter and Mercury; Mercury and Pluto.

14.30. Mercury, Venus, Mars, Jupiter, and Saturn.

14.31. Mercury; Pluto.

14.32.

Planet	Predicted Distance	Actual Distance
Mercury	0.4 + 0.0 = 0.4	0.39
Venus	0.4 + 0.3 = 0.7	0.72
Earth	0.4 + 0.6 = 1.0	1.00
Mars	0.4 + 1.2 = 1.6	1.52
Asteroids	0.4 + 2.4 = 2.8	2.90 (average)
Jupiter	0.4 + 4.8 = 5.2	5.20
Saturn	0.4 + 9.6 = 10.0	9.54
Uranus	0.4 + 19.2 = 19.6	19.18
Neptune	0.4 + 38.4 = 38.8	30.06
Pluto	0.4 + 76.8 = 77.2	39.44

14.33.   15 days.

14.34.   New moon; full moon.

14.35.   No; yes.

14.36.   Yes; no.

14.37.   The tail of a comet is formed by pressure exerted by solar emissions.   Sunlight is partly responsible but the chief influence is the "solar wind" of protons and electrons.   Far from the sun the effect of these emissions is too small to produce a tail.   A comet's tail always points away from the sun since the direction of the solar emissions is always radially outward from the sun.

14.38.   The likelihood is greater because the Martian atmosphere is less dense than that of the earth and hence less effective in causing meteoroids to vaporize before reaching the surface.

Chapter 15

The Universe

THE SUN

The sun is a typical star composed largely of hydrogen with a substantial proportion of helium and small amounts of most of the other elements. The sun's energy originates in the fusion of hydrogen nuclei (protons) to form helium nuclei. A helium nucleus has less mass than the total mass of four protons, and the difference is evolved as energy during the formation of each helium nucleus. The conversion of hydrogen to helium can take place under the conditions believed to exist in the sun's interior in two different ways. In one of them, the *proton-proton cycle,* collisions of protons result in the formation of heavier nuclei whose collisions in turn yield helium nuclei. The *carbon cycle* is a sequence of steps in which carbon nuclei absorb a succession of protons until they ultimately disgorge helium nuclei to become carbon nuclei once more. In the sun the proton-proton cycle predominates; in hotter stars, the carbon cycle predominates.

SOLAR ATMOSPHERE

The light that reaches us from the sun is emitted by a glowing gas envelope called the *photosphere* whose temperature is about 5700 °C. Large flamelike *prominences* frequently extend outward from the solar surface. The *corona,* which is visible during solar eclipses, is a very hot (over 1 million °C in its outer part) cloud of ionized gas that extends for millions of miles outward from the sun. The density of the corona is very low, so despite its high temperature it radiates much less light than the photosphere.

Ions, largely protons and electrons, continuously stream outward in the corona and constitute the *solar wind* that helps cause comet tails to point away from the sun. The faster ions from the sun produce the *aurora* in the upper atmosphere of the earth by exciting the atoms and molecules there to radiate light.

SUNSPOTS

Sunspots are dark markings on the solar surface that persist for a few hours to a week or more. A sunspot appears dark because its temperature is about 1000 °C cooler than the rest of the photosphere. The number of sunspots varies through an 11-year cycle; at sunspot minimum, few if any spots are visible, whereas at sunspot maximum 100 or more spots may be present at one time. It seems possible that a number of terrestrial phenomena are correlated with the sunspot cycle, in particular certain periodic weather variations. Strong magnetic fields are associated with sunspots; because the polarities of these fields are reversed in successive 11-year cycles, the true period of the sunspot cycle is 22 years.

A *solar flare* is a sudden release of a large amount of energy from the solar surface. In a flare, a region of the surface becomes much brighter than usual, and ultraviolet light, X-rays, and streams of fast ions are emitted copiously. Flares occur most often in the neighborhood of sunspot groups and are most frequent at sunspot maximum. The various radiations associated with flares affect the earth's ionosphere and magnetic field, produce exceptional auroral activity, and are dangerous to astronauts.

STARS

Stars vary widely in size and brightness, but they all contain roughly similar amounts of matter. The most common small stars are *white dwarfs,* which are only slightly larger than the earth although they have the mass of the sun. The atoms in a white dwarf have collapsed and their nuclei and electrons are packed closely together. Near the other extreme are *red giants,* some as large across as the orbit of Mars, whose densities are comparable with that of the earth's atmosphere at high altitudes. The sun is an average star in most respects. The color of a star depends upon its surface temperature, as in the case of a poker held in a fire: the hottest stars are blue, those of intermediate temperature are white or yellow, and the coolest are red.

Under certain circumstances atoms emit or absorb light of certain wavelengths that are characteristic of the element involved. Each such wavelength is referred to as a *spectral line,* and from the analysis of the spectrum of a star it is possible to infer the star's composition, state of matter, whether magnetic fields are present, and whether the star is in relative motion toward or away from the earth.

STELLAR EVOLUTION

Stars are believed to originate in gas clouds in space, which consist largely of hydrogen. Local concentrations occur from time to time in such clouds, and if the density of one of them is sufficiently great, gravitation will both attract more gas and cause the accumulation to contract. The contraction liberates potential energy which heats the infant star and causes it to glow, and eventually it becomes hot enough for nuclear reactions to occur that convert its hydrogen to helium. Now the tendency of the star to contract gravitationally is balanced by the pressure of radiation from its hot interior, and it maintains a nearly constant size. The period of stability depends on the star's mass: the greater the mass, the more rapidly it consumes its hydrogen content, and the shorter its period of stability.

When the hydrogen in a star's interior is sufficiently depleted, it contracts, which releases gravitational potential energy and, by raising the temperature near the center, increases the rate of nuclear reactions there. Some of these reactions now involve the building-up of nuclei larger than that of helium, for instance in the combination of three helium nuclei to form a carbon nucleus. The outer part of the star is heated and expands, but the expansion then causes a cooling so the result is a very large, cool star with a hot core. Such a star is a red giant. When the sun becomes a red giant about 5 billion years hence, it will expand until it swallows up some or all of the inner planets.

It seems likely that most stars end their lives as white dwarfs, though the transition from red giant to white dwarf is not well understood. White dwarfs cannot exceed 1.2 solar masses, but a variety of ways are known by which stars more massive than this can eject sufficient matter to become white dwarfs. Another possible destiny for a star is to become a *neutron star,* a sort of superdwarf much smaller and denser than white dwarfs and with such enormous internal pressures that the only stable form of matter is the neutron. Calculations show that a neutron star of one solar mass would have a diameter of only 10 km or so. *Pulsars,* which emit brief, intense bursts of radio waves at regular intervals, are believed to be rotating neutron stars with magnetic fields that lead to radio emission in narrow beams; as a pulsar rotates, its beams swing with it to produce the observed fluctuations.

GALAXIES

The stars are not uniformly distributed in space but occur in aggregates called *galaxies.* Most galaxies are either elliptical or spiral in character. Elliptical galaxies consist of very old stars and range in shape from spheres to fairly flat ellipsoids; the most common galaxies are dwarf elliptical ones that contain only a few million stars.

The stars that make up the Milky Way are part of a spiral galaxy that includes the sun. Most of the stars in our galaxy, as in other spiral galaxies, are concentrated in a relatively thin dislike region which has a thicker central nucleus, much like a fried egg. The disk is about 100,000 light years in diameter and the stars in it are concentrated in two spiral arms that extend from the nucleus. The sun is located about 30,000 light years from the center of the galaxy and, like the other stars in the galaxy, revolves around the center; the sun's period of revolution is about 200 million years. The stars in the spiral arms are of different ages, including very young ones still in the process of formation. Associated with the galaxy are a number of old stars that form a sort of halo or corona around the central disk, so that the true form of the galaxy as a whole is roughly spherical.

THE EXPANDING UNIVERSE

The spectral lines of all galaxies are found to be shifted toward the red by an amount that increases with distance. The only explanation in accord with existing experimental and theoretical knowledge is that the red shifts originate in the Doppler effect and signify motion away from the earth; the proportionality between recession velocity and distance means that the entire universe must be expanding, so that an observer anywhere finds that everything else is moving away from him.

(When there is relative motion between a source of waves and an observer of them, the frequency perceived by the observer is not the same as that produced by the source. Thus the frequency of a fire engine siren appears higher when it is moving toward us than when it is at rest, and lower when it is moving away from us. Such frequency changes constitute the *Doppler effect*.)

The *big-bang* cosmological model, which is widely accepted today, holds that the present universe came into being perhaps 10 billion years ago in a gigantic explosion. As the primeval matter spread out in space, local concentrations formed that became galaxies. The fastest galaxies traveled farthest since the big bang occurred, which accounts for the correlation between red shift and distance. A modification of the big-bang model suggests that gravitation will slow down and eventually reverse the expansion, so that the universe will contract into a single mass again in the distant future. This mass will then explode in another big bang, starting a new expansion. Such an oscillating universe might have a period of 80 billion years for each cycle.

Solved Problems

15.1. Why is it impossible for combustion to be the source of the sun's energy?

Combustion involves the reaction of oxygen with another substance to form an oxygen-containing compound. Combustion cannot occur in the sun because the high temperatures there would prevent the formation of any such compound. Also, even if combustion could occur, the energy liberated would be totally inadequate to account for the observed rate at which the sun radiates energy.

15.2. Is it possible for an object with the mass and composition of the sun to exist without radiating energy?

Such an object must contract owing to gravitational forces. The contraction causes both a rise in temperature and an increase in density, as a result of which the hydrogen present begins to be converted to helium with the release of considerable energy. Thus any object with the mass and composition of the sun must radiate energy like the sun.

15.3. The sun rotates on its axis, as the earth and other planets do. (*a*) State two methods for determining the rate at which the sun rotates. (*b*) Is the rotation rate uniform?

(a)   Sunspots appear to move across the sun's disk because of solar rotation. From the velocity of this motion the rate at which the sun rotates can be found. Another approach is to measure the Doppler shift in the spectral lines of radiation from the limbs (edges) of the sun's disk. At one limb the spectral lines are found to be shifted toward the blue end of the spectrum, which signifies motion toward the observer, and at the other limb the spectral lines are found to be shifted toward the red end of the spectrum, which signifies motion away from the observer. From the amounts of each shift the rotation rate can be determined.

(b)   The period of solar rotation is about 25 days at the equator and increases to about 35 days near the poles.

15.4.   How can the composition of the sun be determined?

The presence of the spectral lines of a particular element in the solar spectrum means that this element must be present in the sun.

15.5.   How is it possible for helium to have been discovered in the sun before it was found on the earth?

Helium is abundant in the sun, and its spectral lines appear as part of the solar spectrum. These spectral lines at first could not be identified as those of any terrestrial element, so it was assumed they corresponded to a new element that was called helium after *helios,* which is Greek for "sun." Later a new gas was discovered on the earth, and its spectrum turned out to be the same as the helium spectrum.

15.6.   What is the evidence for the belief that strong magnetic fields are associated with sunspots?

Spectral lines from atoms in a magnetic field are each found to be split into two or more component lines, with the spacing between the component lines varying with the strength of the field. This phenomenon is called the *Zeeman effect* after its discoverer. The presence of split lines in the spectra of sunspots signifies the presence of a magnetic field, and the strength of the field can be determined from the separation of the component lines.

15.7.   Why are auroras most common near the polar regions?

The earth's magnetic field deflects the streams of solar ions that cause the aurora so that they usually reach the atmosphere in doughnut-shaped zones centered about the north and south geomagnetic poles.

15.8.   Astronomers use the *light year* and the *parsec* as units of distance. How are these units defined?

The light year is the distance traveled by light in one year. Since the velocity of light is 3×10^8 m/s, the light year is equal to 9.46×10^{15}m, which is 5.88×10^{12} miles.

The parsec is the distance from the earth at which a star will change its apparent position in the sky back and forth through an angle of 1″ in the course of a year, during which the earth makes a complete circuit of its orbit.   (1° = 60′ = 60 min and 1′ = 60″ = 60 sec, so 1° = 3600″.)   One parsec = 3.26 light years = 3.086×10^{16} m.

15.9.   The brightness of certain stars fluctuates.   What are the three chief classes of variable stars?

(1) **Eclipsing binaries.**   These consist of double stars of which one member of the pair periodically moves across the face of the other to produce a change in apparent brightness. Strictly speaking, an eclipsing binary is not a true variable since the luminosity of each member of the pair does not change.

(2) **Pulsating variables.**   Such a star expands and contracts in a more or less regular rhythm, and its light output fluctuates in the same rhythm. Apparently a change in temperature accompanying each change in size is primarily responsible for the variation in luminosity, with the change in size a secondary factor. Most pulsating variables are giant or supergiant stars. The *Cepheid variables,* which fluctuate with periods of a few days, are significant because their periods are related to their luminosities. Thus they provide a way to find the distances of remote star groups: from the period of a Cepheid in such a star group the luminosity of the Cepheid can be established, and comparison with its brightness as seen from the earth then yields the distance of the star and so of the group.

(3) **Eruptive variables.**   Most such stars exhibit sudden increases in brightness, notably the *novae* and *supernovae*; a few decrease in brightness. A nova is a small, hot star that abruptly flares up to thousands or tens of thousands of times its normal luminosity as it ejects a shell of gas that expands rapidly. A nova may

take a day or less to attain its maximum brightness, then it gradually declines to its former brightness over a longer period, up to a year or more. A supernova flares up to hundreds of millions of times its normal luminosity, sometimes outshining the entire galaxy of which it is part. In a supernova the ejected material is a substantial part of the original star, much more than in the case of a nova. The *Crab nebula* is the expanding cloud of gas from a supernova observed in 1054. A pulsar is at the center of the nebula, which suggests that during the supernova explosion the core of the original star was imploded sufficiently to produce a neutron star.

15.10. **What is a "black hole"?**

According to the general theory of relativity, light waves are affected by gravitational fields. For example, starlight passing near the sun is deflected to a small but measurable extent by the sun's gravity. The smaller an object and the more massive it is, the stronger the gravitational field at its surface; if this field is strong enough, light will be unable to escape from the object, and it is a "black hole" in space. Conceivably some stars whose mass exceeds 1.2 solar masses ultimately contract to become black holes instead of shedding enough matter to become white dwarfs. (The upper limit to the mass of a white dwarf is 1.2 solar masses.)

15.11. **The earth undergoes four major motions through space. What are they?**

The earth (*a*) rotates on its axis, (*b*) revolves around the sun, (*c*) revolves with the sun around the center of the Milky Way galaxy, and (*d*) moves with the galaxy as a whole in the expansion of the universe.

15.12. **How can the rotation of a spiral galaxy be experimentally determined?**

If the galaxy is so oriented that we see its disk edge-on or nearly so, the stars on one side of the center are moving toward us and those on the other side are moving away from us. These motions produce Doppler shifts in the spectrum of the galaxy that can be detected.

15.13. **Do radio telescopes actually magnify something? If not, why are larger and larger ones being built?**

Radio telescopes are giant antennas connected to sensitive instruments that detect radio waves from space; they do not magnify anything. The larger a radio telescope is, the better able it is to respond to weak radio signals, and the more accurately it can identify the direction from which a given signal is coming.

15.14. **What is a *quasar*?**

A quasar ("quasi-stellar radio source") appears in a telescope as a point of light, as a star does, but it is a far more powerful source of radio waves than any known star. The spectra of quasars show large red shifts; if quasar red shifts are related to their distances in the same way as galactic red shifts are, quasars are not only the most distant objects in the universe but are also giving off energy at prodigious rates. The light and radio outputs of quasars sometimes change in periods of a few weeks, so they cannot be more than a few light weeks across, which makes the mechanism of their energy emission even more difficult to explain. On the other hand, if quasars are actually nearby objects located in our galaxy, the energy problem is less severe, but now the origin of their red shifts has no explanation in current knowledge. The nature of quasars is a major challenge to astrophysics.

15.15. **What are *cosmic rays*?**

High-energy atomic nuclei, largely protons, continually rain down on the earth from space. These *primary cosmic rays* circulate throughout our galaxy, to which they are confined by magnetic fields; a few exceptionally energetic protons probably originate outside the galaxy, since the intergalactic magnetic fields are not strong enough to trap them. It seems likely that primary cosmic rays consist of nuclei originally ejected during supernova explosions which were subsequently accelerated by the same magnetic fields that prevent their escape from the galaxy.

When a primary cosmic ray arrives at the earth's atmosphere, it disrupts atoms in its path to produce a shower of *secondary cosmic rays* which are what reach the surface. The secondaries consist of neutrons and protons, mesons and other unstable particles, and electrons and gamma rays from the decay of the latter.

15.16. **Why is it unlikely that most cosmic rays come from the sun?**

A solar origin would produce a day-night variation in cosmic-ray intensity, but no such variation is observed. Also, no mechanism is known by which the sun could accelerate cosmic-ray primaries to the highest energies they are found to have, though such mechanisms are likely to exist in the galaxy as a whole. Some low-energy primaries, however, probably have a solar origin.

15.17. Is cosmic-ray intensity the same everywhere on the earth?

Cosmic-ray primaries are atomic nuclei and hence are charged particles. The earth's magnetic field is not strong enough to influence the paths of the more energetic primaries by very much, and they arrive equally often everywhere, but the slower ones are deflected so that fewer of them reach the equatorial regions than the polar regions.

15.18. Why is the sun thought to be part of the central disk of our galaxy?

The Milky Way is composed of stars in the spiral arms of our galaxy and so defines its central disk. Since the earth is close to the plane of the Milky Way, the sun must be part of the central disk of the galaxy.

15.19. What is the *steady-state* theory of the universe?

The basis of this theory is the belief that the universe is unchanging in its basic character, so that it has no beginning and no end. In order that the density of galaxies in space remain the same despite their observed spreading apart, the steady-state theory postulates that new galaxies are forming in empty space all the time, with the necessary matter coming into being by a process of spontaneous creation. The theory is attractive philosophically in its conception of an eternal universe, always evolving and yet always the same. However, there are serious objections to the notion of the spontaneous creation of matter, and present experimental evidence is not consistent with various predictions of the steady-state theory.

15.20. The most distant galaxies are very faint in even the largest of today's telescopes and reliable data on them is rare. If adequate such data were available, what information would you look for to decide whether the big-bang, oscillatory, or steady-state theory of the universe is nearest the truth?

(1) **Apparent ages of distant galaxies.** Galaxies form, mature, and age in an evolutionary sequence. The light reaching us from a distant galaxy left it long ago; any information we can obtain about a galaxy say 5 billion light years away represents its physical state 5 billion years ago. According to the big-bang and oscillatory theories, all galaxies came into being at about the same time soon after the expansion of the universe began. Thus these theories predict that distant galaxies will appear younger than nearby galaxies. According to the steady-state theory, nothing fundamental ever changes in the universe, so distant galaxies will appear the same as nearby ones.

(2) **Density of galaxies.** According to the big-bang and oscillatory theories, the galaxies long ago were closer together than they are today. Thus distant galaxies, which we see as they were in the past, should be closer together in space than nearby galaxies, which we see as they were more recently. According to the steady-state theory, the density of galaxies should not vary, since as they move apart new ones come into being.

(3) **Rate of expansion.** The big-bang theory predicts that the universe will continue to expand indefinitely, though its rate will slow down due to gravitation; the oscillatory theory predicts that gravitation will eventually cause the universe to contract; and the steady-state theory predicts that the expansion will continue forever at a constant rate. Thus determining exactly how the rate of expansion varies with distance for the farthest galaxies will help establish which of the three theories is correct—if indeed any of them is correct in its present form.

15.21. If the universe originated in a big bang, in its early moments it must have been a very hot, dense *primeval fireball*. Because of its high temperature, radiation from the fireball would have been in the high-frequency region of the spectrum, chiefly in the form of X-rays. Today space is pervaded by a sea of low-frequency radio waves which have been identified as the remnants of the fireball radiation. (This is one of the pieces of evidence that contradict the steady-state theory.) What caused the change in frequency?

The expansion of the fireball produced Doppler shifts in the radiation that decreased the frequencies present in it, just as the light from distant galaxies is found to exhibit red shifts.

Supplementary Problems

15.22. What aspect of the formation of helium from hydrogen results in the evolution of energy by the process?

15.23. The sun's mass is 2×10^{30} kg, and it currently is losing about 4×10^9 kg of its mass per second as its hydrogen is converted into helium. If the sun has been radiating energy at the same rate as at present during the 4.5-billion-year existence of the earth, what fraction of its original mass has been lost? What does this suggest about the possibility that the solar radiation rate has indeed been approximately constant during most of this period of time, as is suggested by geological evidence?

15.24. Why is the corona of the sun ordinarily not visible? How do we know it exists?

15.25. Why do sunspots appear dark if their temperatures are over 4000 °C?

15.26. Why is the sun considered to be a star?

15.27. Into what kind of star will the sun eventually evolve?

15.28. How can the motion of a star be detected?

15.29. Are stars uniformly distributed in space?

15.30. Where is most of the interstellar gas in our galaxy located? What is its chief constituent?

15.31. Cosmic-ray primaries are mostly protons, but few protons are in the cosmic rays that reach the earth's surface. Why?

15.32. What effect, if any, would the disappearance of the earth's magnetic field have on the distribution of cosmic rays around the earth?

15.33. According to current ideas, where are elements heavier than hydrogen formed?

15.34. The spectra of quasars exhibit red shifts, never blue shifts. Why does this suggest that quasars are not members of our galaxy?

Answers to Supplementary Problems

15.22. A helium nucleus has less mass than the total mass of the four hydrogen nuclei that combine to form it, and the "missing" mass appears as energy.

15.23. The fraction of the sun's mass that has been lost in this period if its radiation rate has been unchanged is 2.8×10^{-4}, which is 0.028%. So little of the sun's mass has been converted to energy since the earth was formed that it is entirely possible that the sun's radiation rate has not changed appreciably during this period.

15.24. The photosphere is so much brighter than the corona that it cannot be seen unless the photosphere is masked, which is done by the moon during a total solar eclipse and also in special telescopes called *coronagraphs*.

15.25. Sunspots appear dark only by comparison with the rest of the photosphere, whose temperature is higher.

15.26. The sun's properties, such as size, mass, temperature, composition, and luminosity, are those of a typical star.

15.27. The sun will eventually become a white dwarf.

15.28. A star moving toward or away from the earth will show a Doppler shift in its spectrum toward the blue or the red end respectively. A star moving across our line of sight will change its position relative to other stars, as revealed by photographs taken at different times.

15.29.   Stars are concentrated in galaxies that are relatively far apart from one another.

15.30.   Most of the interstellar gas is located in the spiral arms of the galaxy, and its chief constituent is hydrogen.

15.31.   Cosmic-ray primaries lose their initial energies in collisions with the nuclei of atoms in the atmosphere, and the cosmic rays that reach the earth are almost all secondaries produced in these collisions.

15.32.   In the absence of the geomagnetic field, cosmic rays would arrive at the same rate everywhere on the earth.

15.33.   The elements are formed in the interiors of stars.

15.34.   If quasars were members of our galaxy, at least some of them would have components of motion toward the earth and would accordingly have Doppler shifts in their spectral lines toward the blue end of the spectrum.

Index

Index

Catalog

If you are interested in a list of SCHAUM'S
OUTLINE SERIES in Science, Mathematics,
Engineering and other subjects, send your name
and address, requesting your free catalog, to:

SCHAUM'S OUTLINE SERIES, Dept. C
McGRAW-HILL BOOK COMPANY
1221 Avenue of Americas
New York, N.Y. 10020